T0211387

Smart Ports

Weijian Mi · Yuan Liu

Smart Ports

Weijian Mi
Shanghai Maritime University
Shanghai, China

Yuan Liu
Shanghai Maritime University
Shanghai, China

ISBN 978-981-16-9891-0   ISBN 978-981-16-9889-7   (eBook)
https://doi.org/10.1007/978-981-16-9889-7

Jointly published with Shanghai Scientific and Technical Publishers, Shanghai, China
The print edition is not for sale in China Mainland. Customers from China Mainland please order the
print book from: Shanghai Scientific and Technical Publishers.

Translation from the Chinese Simplified language edition: "Zhihui Gangkou Gailun/智慧港口概论" by
Weijian Mi and Yuan Liu, © Shanghai Scientific and Technical Publishers 2020. Published by Shanghai
Scientific and Technical Publishers. All Rights Reserved.

This Springer imprint is published by the registered company Springer Nature Singapore Pte Ltd.
The registered company address is: 152 Beach Road, #21-01/04 Gateway East, Singapore 189721,
Singapore

Preface

Since the world's first automated container terminal was officially operated in Rotterdam in the Netherlands in 1993, port intellectualization has been one of the focuses of academia and industry. Since the beginning of twenty-first century, with the rapid development of high and new technologies represented by communication technology, network technology, and intelligent technology, the number of automated container terminal constructed has increased. In recent 10 years, China's port intellectualization level has also made great progress. The construction of automated container terminals has been accelerated significantly, and the construction of intelligent dry bulk terminals has also started trial operation. In the process of promoting One Belt and One Road Initiative and the Maritime Silk Road in China, the construction of smart ports has increasingly become a hot spot in the development and construction of ports in the world.

The new concept of smart port around the intellectualization of port and construction of smart port is put forward in this book, namely, with the cyber-physical system as the structural framework, through the innovative application of high and new technology, the logistics supply side and the demand side are integrated into the collection-distribution-transportation system, which has greatly improved the comprehensive information processing capacity of the port and its related logistics park and the optimal allocation capacity of related resources, to build the new ecological port with intelligent supervision, intelligent service, autonomous loading and unloading as its main form, and capable of providing modern logistics industry with high-safety, high-efficiency and high-quality services. Besides, it is also proposed to establish an innovative, coordinated, green, open, and shared development ecology in the book and it is illustrated that the essence of smart port ecological system is innovation, which is also the premise and basis for the sustainable and well development of smart port.

The most important high-tech technologies in the construction of smart port are introduced in the book from aspects of concept description, development condition, and application examples, including cyber-physical system, digital middle-office, blockchain, artificial intelligence, machine vision, AR/VR technology, simulation and emulation analysis, digital monitoring and diagnosis, etc., which constitute the main technical foundation for the intelligent transformation and development of traditional terminals and the construction of new automated terminals in the future. In the end, the future development trend and target of smart port is summarized.

I hope that this book will provide some useful examples and guidance for managers, technicians, and researchers who are engaged in the construction of smart ports and who are already working in this field.

Shanghai, China   Weijian Mi

Acknowledgments Most of the cases and results of scientific research in the book are the accumulation of the author's team work in the research of intelligent technologies in the port for a long time, as well as practice experience from cooperation with relevant port enterprises in port construction. Preparation of the manuscripts for this book has involved efforts and comments of my colleagues, and appreciation goes to Dr. Fan Shu, Dr. Hongxia Jia, Engineer Langjie Shen, Dr. Ning Zhao, Dr. Nannan Yan, Dr. Yifan Shen, Engineer Zhiwei Zhang, Dr. Xiaoming Yang, Dr. Haiwei Liu, Dr. Yang Shen, Dr. Mengjue Xia, etc. Meanwhile, I would like to express my heartfelt thanks to my colleagues and friends who have provided valuable materials, cases, and achievements in this book.

Contents

About the Authors

Weijian Mi received his Ph.D. in Mechanical Engineering from Tongji University. He is now a Full Professor of College of Logistics Engineering, Shanghai Maritime University. He now serves as Executive Director of China Logistics Association, Director of Fault Diagnosis Technology Committee of China, Executive Director of China Mechanical Engineering Society, Secretary-General of Testing and Controlling branch of China Construction Machinery Association, Vice Chairman of Shanghai Vibration Engineering Society, Associate Managing Editor of Chinese Journal of Construction Machinery and Journal of Shanghai Maritime University. He directs a research team working on Port Intelligence and Port Machinery Maintenance. His current research interests include terminal operation and resource optimization, port logistics simulation optimization and port machinery fault diagnosis. He has led his research team to win the National Award for Progress in Science and Technology (second class), Golden Award of Next Generation Automated Container Terminal Challenge in Singapore, and the Shanghai Award for Progress in Science and Technology (second class).

Yuan Liu from Shanghai Maritime University, China, received her Ph.D. from Tongji University, China. She has studied as Visiting Scholar at Vienna University of Technology, Austria, and taught Manufacturing Processes at Saint Martine's University, US. Now she mainly teaches Statics and Mechanics of Materials, Engineering Design Methodology, Manufacturing Processes, etc. Her research interests are rolling contact fatigue between rail and wheel, safety evaluation and management of metal structures of large port machinery, intelligent logistics of ports, etc. She has published eight papers as the first author in *Journal of Applied Science, Engineering Mechanics, Journal of Coastal Research*, and other domestic and foreign journals, among which one paper was included in SCI and three in EI. Her industrial experience includes work and research in metal structure tests of cranes in Tianjin Port, Shanghai Port, Zhanjiang Port, etc. in China.

Chapter 1
General Introduction

1.1 General Introduction to Smart Port

What is Smart Port? Why we have to build smart ports and how? First of all, ports are a social and economic existence with a long history. They are links between "water" and "city", and almost all the cities are built by the water at all times all over the world. With the development of society and economy, ports are not only the distribution center and logistics hub of domestic and foreign trades, but also the barrier and key node of national social and economic security. Various industrial zones and cities relying on port development follow a same development path of "promoting port with city and invigorating city with port". Smartness falls into another social, economic and cultural category. Traditional culture regards it as a symbol of intelligence and wisdom, which is reflected in the context of Buddhism as a better ability to understand and solve problems. In the modern sense, smartness represents an innovative, coordinated, green, open and shared development ecology. The concept of Smart Port and its construction reflect an inevitable trend of historical development. Meanwhile, Smart Port extends the functional concept of traditional port to waterless port, logistics zone, bonded zone and free trade zone.

To be specific, the definition of Smart Port is as: a system with cyber-physical systems as the structural framework, in which logistics suppliers and demanders are involved in the integrated system of collection, distribution and transportation through innovative application of high and new technologies. It has greatly enhanced the port and its related logistics zones' ability to process information comprehensively and optimized the allocation of related resources. Intelligent supervision, intelligent service and autonomous handling have become the main forms of the new ecological port, which can provide services of high safety, high efficiency and high quality for modern logistics industry. It can be seen from the definition of Smart Port that the Internet, mobile Internet, Internet of Things (IOT) and industrial Internet constitute the most important infrastructure of Smart Port. High and new technologies, including 5G, cloud computing, big data, artificial intelligence, blockchain, etc., are perfectly integrated with port functions. In the traditional sense, logistics service

W. Mi and Y. Liu, *Smart Ports*, https://doi.org/10.1007/978-981-16-9889-7_1

providers such as ports, cargo carriers, shipping agents and freight forwarders are all one-way suppliers to provide services to the demanders of logistics. With the diversification of service and the convenience of logistics information, the demand for logistics services becomes more and more complicated. In order to improve the efficiency and quality of logistics, it is urgent for demanders of logistics to participate in and involved in the integrated logistics system of collection and distribution.

For port logistics process, due to the innovative application of high and new technologies, various information is gathered in the port, multifarious advanced automation equipment introduced to the operation process, while electronization of logistics documents becomes increasingly prevalent, and financial payment means innovative. The smart port promotes the reengineering of port logistics process, from star structure of traditional logistics service taking the port as hub node to network architecture linking various logistics chain, which provides technical possibilities for intelligent supervision, intelligent service, and autonomous handling. From the perspective of function, the composition of the smart port can be shown in Fig. 1.1.

Taking "to be innovative, coordinative, green, open, sharing" as development concept, "to be safer, more comfortable, more environment-friendly, more efficient" as development goals, and "sustainable development driven by innovation" as development mode, smart ports are sure to become the main pillar supporting intelligent traffic and smart city.

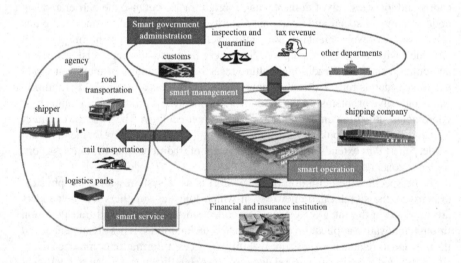

Fig. 1.1 Functional composition of smart port

1.2   Smart Port and Revolution of New Technologies

The basic concept of Smart Port is originated from Smart Factory, which stemmed from the revolution of new technologies and often associated with Industry 4.0. Now, the ubiquitous application of new technologies led by artificial intelligence based on cyber-physical system has opened a new era of the industrial revolution, referred to as Industry 4.0. Industry 4.0 can be narrowly defined as "Smart Factory, Smart Manufacturing", from which "Smart Port, Smart Handling" can be derived.

The first smart factory in the world, the Siemens Electronics Manufacturing plant in Amberg (shown in Fig. 1.2), Germany, was founded in 1989. This factory can produce about 15 million pieces of Siemens products every year. According to the annual production period of 230 days, it can produce one control equipment at each second on average. Among every 1 million pieces of products, there are about 15 defective products, with reliability reaching 99% and traceability reaching 100%.

In this smart factory, people, machines and resources can "communicate" with each other. Smart products "know" the details of how they are manufactured and what they will be used for. They will actively answer questions about the manufacturing process, such as "when am I made?" "What parameters should be used to process me?" "Where should I be transported?" The realization of these functions depends on the cyber-physical system, and the whole smart factory has three main systems of intelligent technologies. The first is the digital twin system for design, which can complete the simulation of manufacturing during the design phase of the product; the second is the real-time dynamic and automatic logistics distribution system, which

Fig. 1.2 The smart manufacturing workshop in the Siemens Electronics Manufacturing plant in Amberg

is capable of ensuring the timely, accurate, automatic distribution of manufacturing logistics to the specific station; the third is the real-time dynamic control system of intelligent manufacturing to realize the flexible manufacturing of mixed production lines. The three systems work together via the cyber-physical system, which can fully realize the automatic intelligent manufacturing of products in the smart factory.

Compared with Smart Factory, the realization of Smart Port is more difficult. In the smart factory, the intelligent manufacturing of products is finally completed through the automated assembly lines, while the control beat of the station and the flow direction of products are relatively fixed. However, the process of handling and transportation of port logistics is more complex, and the uncertainty of logistics information is relatively high, so it is liable to be restricted by many conditions of environment and climate. In addition to control in Kinematics, the handling and transportation process of port logistics also needs to realize control in Dynamics. The most important function of intelligent handling of the smart port has higher requirements on specialized technologies and their combination. Therefore, the advent of the smart port, the counterpart of the smart factory, is relatively late. At present, the rapid development of high and new technologies will undoubtedly provide powerful technical support for the construction and development of smart ports. Among them, edge calculation has effectively solved the connection and information-based control between equipment and equipment, equipment and external physical entity to the end and industrial system, which has been widely used to realize the accurate positioning of vehicles, spreaders and containers. Machine vision has been used for the identification of container no., invasion of objects or personnel on the route, keyhole location, etc. 5G has provided a mobile digital communication network with large capacity, low latency and high speed, making remote control and automatic monitoring of devices possible. Digital middle-office system and its cloud computing have integrated information from different physical fields and various applications of different service objects into a unified platform to form innovative service and innovative ecology. Big data and artificial intelligence organically have combined human experience and knowledge as well as dynamic information of various business processes to form a new format of port logistics. Without the successful application of high and new technologies, the vigorous development of smart ports will not be possible.

1.3   Development History of Smart Port

Looking back at the development history of Smart Port, people's understanding of Smart Port has been deepening and expanding. In general context, Smart Port can be understood in narrow sense and in broad sense. The narrow sense of Smart Port mainly revolves around intelligent handling and focuses on autonomous handling. In the broad sense, the port is taken as the hub, the intellectualization of the whole logistics chain is the core, with intelligent control and intelligent service becoming the main goals. Therefore, if the automated container terminal is regarded as the

Fig. 1.3 ECT in Rotterdam, the Netherlands

beginning of the smart port development, the year of 1993 can be seen as the first
year of the construction of the smart port, when the official operation of the world's
first automated container terminal (Fig. 1.3) was initiated in ECT Delta Sealand in
Rotterdam, the Netherlands. However, the construction of smart ports around the
world developed slowly in quite a long time, the main reason of which is related to
the technical performance.

In 2017, Yangshan (Phase IV), the world's largest single automated container
terminal, was put into operation, which marked that the development of Smart Port
had switched into the fast lane, the overall infrastructures and facilities as well as
software and hardware conditions reached the level matching with the smart port.
All kinds of new technologies and systems developed from automated terminals
can be applied to the intelligent transformation of traditional terminals, such as
remote control technology, visual identification technology, wireless transmission
and communication technology of signals in large capacity, real-time positioning
technology with high precision, remote digital diagnosis technology for equipment,
real-time scheduling system for handling facilities, real-time scheduling system for
horizontal transportation, intelligent TOS, simulation and emulation system for operation, etc. With the rapid development of "new infrastructure construction" in China,
the external conditions for the development of Smart Port have been increasingly
optimized, while the inner construction of the smart port has been further improved.
Looking at the whole world, in addition to the rapid development of automated terminals, especially for the automated container terminals, major ports in the world have

formed their own development approaches, advantages and innovation achievements
of Smart Port according to their own conditions and characteristics.

Taking PSA (Pasir Panjang Terminal) in Singapore as an example. In 1997, a
remotely-operated-and-controlled overhead crane system was built to realize the
semi-automated operation of the storage yard, as shown in Fig. 1.4. In 2014, the
creative design of the next generation of container terminal was put forward, as
shown in Fig. 1.5, introduced the concept of 3D port of configuring terminal layout
with multilayer structure, and facilitating bonded logistics and trade services with
seamless connection to logistics parks. In the port strategy for the year of 2030
of Singapore, it has been further proposed that to be efficient, intelligent, safe and

Fig. 1.4 The semi-automated storage yard of PSA in Singapore

Fig. 1.5 The creative design of the next generation of container terminal in Singapore

green are the four directions of development, realizing the smart port with intelligent operations, and making it an important part of smart city of Singapore.

In 2016, Shanghai Port put forward a new concept of "Smart Port Driving Future Trade", to actively explore the direction of intelligent transformation and upgrading through differentiated value proposition and competitive advantages, including:

①   **Facility operation automation**. Through standardization and optimization of the operation processes, manual participation can be reduced to achieve unmanned operation, and the automation and remote auxiliary control and operation of the mechanical facilities in the yard be realized, to improve the operation efficiency and accuracy.

②   **Intelligent scheduling in port**. The seamless connection between instructions from information system and control functions of facilities in the terminal can be realized with the help of ICT technology, system engineering and artificial intelligence and other achievements. Varieties of transportation resources can be allocated and scheduled to the most effective and most reasonable degree, according to different operating conditions and environments, so as to ensure the smooth port logistics and efficient operation of the terminal.

③   **Visualization of information data**. As the information integration center, the smart port will focus on the acquisition, control and processing of information. With the cyber-physical system, an efficient and integrated digital presentation and real-time interaction of the operating environment inside and outside the port will be formed. Meanwhile, the information interconnection also promotes the development of the logistics platform of smart port. The port operations are seamlessly connected with external logistics services, and information resources during each step of maritime shipping are fully converged and timely shared to avoid the long-term existed isolated island of information of various business links, which makes it possible for the business of each step to operate in coordination. It also meets the user's demand for complete, accurate, timely and convenient information acquisition, and improves the port's ability to provide information resources as well as service quality for the logistics demander. The construction of Yangshan automated terminal-Phase IV (see Fig. 1.6) fully reflects these new concepts.

In recent years, based on the general principle and goal of "to be technology-led, innovation-driven", the smart port construction of Tianjin Port has been focusing on the three core businesses of smart operation, smart trade and logistics, and smart ecosystem, and the integration of various businesses has been promoted with a unified smart port portal. Smart operation is mainly embodied in improving efficiency and security of port operations through intelligent facilities and IOT technology, including the application of automated terminal and intelligent yards, autonomous container trucks, intelligent tally, intelligent gates, intelligent traffic guidance, etc., as shown in Fig. 1.7. Smart trade and logistics is reflected in the integration of information flow and port services, the construction of comprehensive logistics and trade platform to facilitate service, including establishing diversified port information service system, optimizing the environment of port customs clearance, combining smart

Fig. 1.6 Yangshan automated terminal-Phase IV

Fig. 1.7 The automated terminal of Container Terminal Co., Ltd. (Northern Area) of Tianjin Port

logistics, commercial trade, data integration, supply chain services in key areas, so as to improve customer experience and customer satisfaction. Smart ecosystem focuses on the extension of service functions and expansion of the ecosystem through the Internet construction to cultivate innovative ability in business model, including strengthening cooperation with domestic and foreign shipping companies, shippers, traders, railway and highway departments, transportation enterprises, port authorities, financial institutions, etc., to create symbiotic and sharing ecosystem for the port, form the new development pattern of port and city integration, and build a green, smart, hub port of world-class.

1.4 Current Construction of Smart Port

At present, Smart Port has made rapid progress in the construction and development of port hubs. In China, the first automated container terminal was built in 2014 and the world's largest automated container terminal put into operation in 2017, which fully reflects the development speed of China. Table 1.1 shows the construction of representative automated container terminals in the world since 2010. The intelligent transformation of traditional container terminals has reached a high technical level in technical fields such as automatic identification of facilities, autonomous driving, autonomous handling, remote operation and control, etc. Especially in the development and application of intelligent control software, including intelligent TOS, intelligent stowage, intelligent container collection, intelligent ship unloading, intelligent ship control, intelligent storage yard, etc., it has reached the world advanced level.

So far, the general layout and handling technology of automated container terminals in the world are basically the same. The operations of quay cranes adopted at the quayside are fully automatic or semi-automatic via remote control. Horizontal transportation is mostly fulfilled by AGVs, and the application of autonomous container trucks starts trial operation. Autonomous handling is realized at the ends of the storage yard, and the yard is arranged perpendicular to the quay (see Fig. 1.8). The biggest advantage of the automated container terminal of such mode is that the container is handled at the front and rear ends, and the internal and external container trucks are effectively separated. The operation is safe and reliable, the handling location is relatively fixed, the control difficulty is reduced, and there is mature experience for reference. The main disadvantage is that the yard crane has undertaken almost half of the horizontal transportation tasks. Meanwhile, the number of container handling and handover times has been increased by more than 40%, which has greatly affected the operation efficiency.

With the mature application of iAGV and autonomous container truck in China, an automated container terminal mode with the storage blocks configured parallel to the quay and the handling operation by the side of the yard (see Fig. 1.9) has been proposed and put into construction. The handling mode of such automated yard can be regarded as the automated twin of the mode in the manual container terminal yard. Via the application of the double-cantilever-beamed yard crane, the utilization rate of the yard has been increased, the operation flexibility enhanced, the horizontal transportation efficiency improved, and the operation process of the whole terminal is smoother. Although the requirements on operation control are high, such problems can be solved with the current detection and control technologies. The main drawback is that the internal and external container trucks are mixed during operation. It is expected that with the rapid progress of intelligent driving, such as iAGV and autonomous container truck, this problem can be well settled. Such terminal layout has well unified the operations of traditional container terminals, semi-automated container terminals and automated container terminals in form, which is more liable

Table 1.1 The construction of representative automated container terminals since 2010

Automated container terminal	Area (hectare)	Shoreline (meter)	Quay cranes (piece)	Horizontal transportation facilities	Yard cranes
APMT, Virginia, US	93	1000	6 (Phase I)	18 straddle carriers (Phase I)	30 ARMGs (2 for each block, Phase I)
DPW, Antwerp, Belgium	126	1720	9 (Phase I)	47 straddle carriers (Phase I)	14 ARMGs (2 for each block, Phase I)
BEST, Barcelona, Italy	100	1500	18 (Phase I)	42 straddle carriers (Phase I)	80 ARMGs (2 for each block, Phase I)
BNCT, Busan, South Korea	84	1400	8 (Phase I)	20 straddle carriers (Phase I)	38 ARMGs (2 for each block, Phase I)
Khalifa, Abu Dhabi, UAE	90	2400	6 (Phase I)	Straddle carriers (Phase I)	32 ARMGs (2 for each block, Phase I)
Gateway, London, UK	300	2700	8 (Phase I)	28 automated straddle carriers (Phase I)	40 ARMGs (2 for each block, Phase I)
GCT, New York, US	70	800	10	Straddle carriers	20 ARMGs (2 for each block)
DPW, Brisbane, Australia	36	900	4 (Phase I)	Straddle carriers (Phase I)	16 ARMGs (2 for each block, Phase I)
World Gateway, Rotterdam, the Netherland	108	1700	14	59 Lift-AGV	32 ARMGs and 16 C-ARMGs (2 for each block)
APMT MVII, Rotterdam, the Netherland	167	2800	8 (Phase I)	37 Lift-AGV (Phase I)	36 ARMGs (2 for each block, Phase I)
LBCT, Long Beach, US	120	4200	14	93 AGV	70 ARMGs (2 for each block)

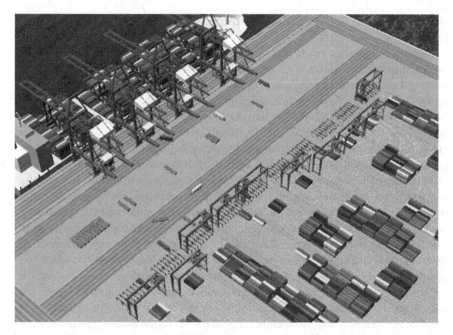

Fig. 1.8 The storage yard is arranged perpendicular to the quay

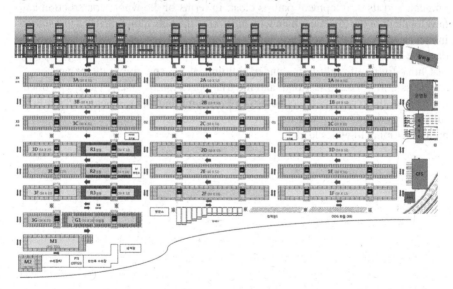

Fig. 1.9 The storage yard is arranged parallel to the quay

to apply intelligent TOS (iTOS) and improve the overall intelligent level of container terminals.

Another reasonable way to promote the development of Smart Port is to proceed the intelligent transformation of the traditional manual terminal according to the existing facilities and technologies. Successful cases have shown that autonomous control of container handling operations in the yard can be realized through automatic transformation of the yard by introducing automatically controlled yard cranes, supplemented with necessary auxiliary manual remote control. Such yard automation mode can not only meet the requirement of seamless connection of the automatic horizontal transportation vehicles, but also adapt to the handover operation mode with the existing manually driven vehicles. One-to-many remote-control mode is generally adopted for the cranes at the quay. Terrain-matching operation can be realized during container handling process and the container position in ship can be automatically guided and confirmed, with the contour scanning of the containers stowed in the ship. The traditional container terminal has broken through the efficiency bottleneck of the traditional handling technologies through intelligent transformation, which has greatly reduced the pressure from the continuously rising labor cost, improved the operation comfort and friendliness of environment, and significantly increased the utilization rate and reliability of facilities. While improving the overall efficiency of terminal operation, it has also greatly reduced the operation cost of terminal.

Smart Port has achieved obvious results in the current port construction and development, and its development path is clear. In terms of the whole construction and development of the logistics chain with the port as the hub, although it has just started in recent years, it has developed rapidly and made rapid progress with each passing day due to the rapid development of high and new technologies. The prosperity of China–Europe freight trains under multimodal transportation, the expansion of waterless ports in the Silk Road, and the trial operation of several electronic documents with the blockchain technology are all symbolic achievements, which have fully shown that China is transforming from an international logistics participant to an international logistics leader.

Chapter 2
Ecology of Smart Port

2.1 Ecological Environment of Smart Port

The original meaning of ecology refers to the unified entity of creatures and environment within a certain space in nature, within which creatures and environment are interacted and inter-conditioned, forming a good dynamic equilibrium. Ecology of smart port can be understood as a value-creating development environment driven by innovation, which is formed around the business of port logistics.

The importance of ecology of smart port is due to the fact that the operation of the world logistics system is accomplished mainly by water transportation. In the process of logistics operation, logistics network is formed by points, lines, planes and corresponding logistics service enterprises, and ports are definitely at the cores and hubs of the logistics chain, which have largely determined the characteristics of high safety, high efficiency and high-quality service of the whole logistics system. Enterprises fulfilling the port functions are the main members of the ecosystem of smart port as well as the leaders and main driving powers of innovation and development. With the advance of Belt and Road Initiative, the development of multimodal transportation and the construction of all kinds of waterless ports, the ecosystem is the premise and foundation to guarantee the sustainable and well development of smart port.

The essence of the ecosystem of smart port is innovation, and this innovative system can be understood as a complicated system, with the port logistics enterprises as the core, and the service enterprises and demanders at each node of the modern logistics chain network as alliance, relying on the government, universities, research institutions, financial intermediary service agencies as the carriers of the ecological elements. Via the effective collaboration between organizations and thorough integration of innovation elements such as manpower, capital, technology, information, the effective convergence of innovation factors can be achieved, bringing value creation and promotion to all kinds of enterprises within the system, realizing the sustainable development of each entity.

© Shanghai Scientific and Technical Publishers and Springer Nature
Singapore Pte Ltd. 2022
W. Mi and Y. Liu, *Smart Ports*, https://doi.org/10.1007/978-981-16-9889-7_2

Therefore, the formation, development and construction of smart port ecology include the following aspects.

First of all, in the process of formation, development and construction of the ecosystem of smart port, the government as the promoter of national technological and institutional innovation, shall play the functions of macro planning, regulation making, policy guiding, finance supporting, service assuring, etc., and provide excellent environments of policy, resource, legislation to support and promote innovative activities within the ecosystem of smart port. For example, the established collaborative innovation centers at all levels are constantly promoting and influencing the main force of innovation and innovative activities in the ecosystem.

Second, the port logistics enterprises, as the main implementer of smart port innovation, are at the core of the ecosystem. Meanwhile, they directly or indirectly connect to all other enterprises in logistics network, including logistics service enterprises, technical supporting enterprises, logistics demanders and related auxiliary enterprises, in solid logistics, capital, information and other contacts. The innovation of each entity in the logistics chain or logistics network can be regarded as an organic component of the ecosystem construction of smart port. In the process of ecosystem construction, the innovation functions, innovation abilities and innovation requirements of the main enterprises and other component enterprises will also have an impact on the relevant decisions of the government in the fields of policy, revenue, finance and legislation.

Third, as the main force of original innovation, universities and research institutes are the source of talent flow and technology flow in the innovation ecosystem. Universities can be directly involved in the creation, development, promotion and application of new knowledge and technologies, presenting strong "Spillover Effect". Universities can also provide a source of innovation for the smart port ecosystem and can be seen as a major supplier of knowledge, technologies and talents. Meanwhile, universities can directly contact the enterprises, as well as influence them via the talent markets, engineering centers, technology transferring platforms and technology markets. Research institutes have similar functions to universities in terms of technology innovation, and can be regarded as the main force for research and development of cutting-edge technologies and specialized technologies.

Fourth, as the main force of technological innovation services, the third-party institutions can provide a large number of socialized and specialized technological consulting services for the main force of innovation, playing an obvious role in communication and integration, and can especially promote the popularization of innovative knowledge and technologies as well as transformation of scientific and technological achievements. Intermediary service organizations such as technology transferring centers, incubation centers, science and technology consultation and evaluation institutions, and industrial technology associations can also promote the innovation and development of enterprises and help build a complete smart port ecosystem.

Fifth, financial institutions, as one of the main investors to innovation, are the important sources of innovation funds in the smart port innovation ecosystem. As an important part of the innovation ecosystem, financial institutions can provide necessary funds and materials for the innovation ecosystem. Meanwhile, under the guidance of high-tech development strategies and industrial policies, governments at all levels will also provide necessary R&D funds and policy-based supporting funds for the innovation ecosystem, which can be regarded as the basis and guarantee for the efficient operation of the innovation ecosystem.

2.2 Ecological Features of Smart Port

In general, smart ports can be described and characterized from five aspects: comprehensive perception, intelligent decision-making, autonomous handling, whole-process participation and continuous innovation. These five features exist at all levels of smart port, reflected in various elements of human, machine and environment, and affect all aspects of the development of smart port.

2.2.1 Comprehensive Perception

From the perspective of engineering psychology, perception is a series of processes of perceiving, feeling, noticing and understanding of consciousness to the internal and external information. Perception can be divided into two processes: sensation and cognition. The perceived information in the sensation process includes the internal physiological condition and psychological activities of the organism, as well as the existence and relationship information of the external environment. Sensation is not only to receive information, but also influenced by psychological effects. In the process of cognition, sensation information is processed in an organized way and the existence form of things is understood and recognized. Based on this definition, the comprehensive perception ability of smart port is to obtain all kinds of information in the ecological environment of smart port accurately, timely and comprehensively, and get the correct response through an effective mechanism. Obviously, comprehensive perception requires not only the ability of perceiving, that is, the ability to be directly aware of various changes of internal factors and external conditions affecting the operation of smart port, but also the ability similar to human perception, that is, the ability to explore and integrate the information obtained by direct sensation from the general and application requirements.

In fact, comprehensive perception is the foundation of all deep intelligent applications, and all the perceptual information may be in a large category, coming from different business levels and different physical domains. Taking the intelligent handling in the automated yard as an example, the perceptual information includes: container and cargo information from optical sensor scanning QR code on electronic document, container no. and vehicle license plate no. from visual sensor scanning the container and vehicle, and storage location of the target container in the yard sent to the driver at the intelligent gate, vehicle-mounted RFID information read at the entrance of the intelligent yard, container and vehicle position information detected by the target detection system (TDS) of the operating yard crane, spreader position and posture information detected by the spreader detection system (SDS), the deep information perceived by the combination of TDS and SDS information to complete the container automatic capture and handling, and the final container location in the yard fed back to the TOS. A handling activity requires so much external information to be perceptively acquired. In order to improve the efficiency of handling operation and the utilization rate of storage yard, it is necessary to perceive and analyze the content of each event and activity of each handling operation process as well as its implied information, which requires data mining from the database of the terminal management system, and this can also be regarded as a deep perception.

Based on the specific requirements of intelligent management, control and service in the terminal, comprehensive perception can transmit the perceived information and data through the cyber-physical system or the industrial Internet (such as the Internet, IOT and Mobile Internet) to the digital middle-office or cloud platform, using all kinds of intelligent processing and intelligent computing technologies to analyze and process as well as integrate and manage the huge amounts of converged data and information (by screening and mining, quality control, standardization, and data integration) to serve the goal of intelligent handling for smart port. Comprehensive perception constitutes the first feature of smart port ecosystem.

2.2.2 Intelligent Decision-Making

Decision-making is an activity that often happens in enterprise management, and scientific decision-making is the core of modern management, running through the whole management activity. In order to achieve a specific goal, the decision makers make the decision and plan for the future action according to the objective possibility and information from the comprehensive perception, after analysis, calculation and judgment of the various factors affecting the realization of the goal on the basis of combining their own thinking, will and experience. Intelligent decision-making in the development of smart port construction is to clarify the decision goal and constraint conditions on the basis of perceptually collecting fundamental decision information, and apply scientific theories, methods and tools to make efficient and effective decisions on issues such as general planning, complex planning and dynamic scheduling.

Intelligent decision-making can be divided into strategic decision-making, battle decision-making, tactical decision-making and behavioral decision-making in different levels, directions and fields. For automated yard of container terminal, for example, if the principles, plans and implementation of general storage in the yard are seen as strategic decisions, then periodic and dynamic storage plans by routes and ports and their implementation can be seen as battle decisions, while specific dynamic plan at each block as well as the corresponding facility scheduling may be regarded as tactical decisions, and automatic control actions during specific container handling process can be seen as behavioral decisions. The purpose of intelligent decision classification is to make more effective use of the perceived information, utilize more appropriately the artificial intelligence knowledge, and respond more quickly to problems to put forward correct decisions. The specific implementation process of intelligent decision-making can be divided into three stages. The first is to formulate the decision problem, that is, with the perceived information and existing experience knowledge, the whole process of the event is clarified, the problem and its key points are determined, and the decision goal is put forward. The second is to diagnose the decision problem, that is to study the general principles and methods, analyze and formulate a variety of alternative plans and measures, predict the possible scenarios and put forward the corresponding countermeasures. The last is to select action plan, that is, the best plan is screened out from all alternatives, and the corresponding feedback system is established. At these three stages, the abilities of intelligent decision-making to formulate problem, predict behavior and make decision should be fully reflected.

In the construction and development of smart port, intelligent decision-making is mainly faced with non-procedural decision-making, that is, the decision-making of novel, complex and uncertain problems in management. There is always no conventional reference for such decision making, and even though it can refer to past experience and similar practices, restudy in light of new circumstances shall be proceeded to make decision. To a large extent, such decision depends on the intelligence and experience of the decision maker in politics, economy and technology, and needs to apply innovative theories, methods and technical means to implement. Figure 2.1 shows the general technical architecture for intelligent decision-making.

Decision-making is an essential step before any goal-directed activity occurs. Scientific decision-making is the key to the success of management, and it is also the main responsibility of modern managers. Construction and development of smart

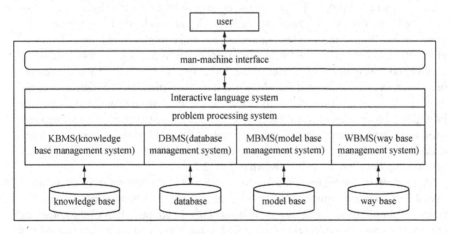

Fig. 2.1 The general technical architecture for intelligent decision-making

port are innovative processes and activities of a complex system, and the constitution of main decision-makers is complex and changeful. Information perceived in decision-making activities has increased greatly, frequency of decision-making activities has been raised, and the cycle shortened, decision-making system has been growing in complexity. Therefore, the universalization and validity in application of intelligent decision-making are undoubtedly important features reflecting the merits of ecological environment of smart port.

2.2.3 Autonomous Handling

Autonomous handling is a technology with which the environment can be independently sensed and automatic control can be realized in handling activities and processes with little or no human intervention. Autonomous handling has combined the perception ability, decision-making ability, coordination ability and action ability of the handling control system. It is the ability to make self-decision according to certain control strategies and continuously execute a series of control functions to complete the expected handling task in the unstructured environment. Since the middle of the twentieth century, human beings have placed great expectations on the construction of intelligent perception and control systems. Intelligence itself is a broad issue, even for artificial intelligence. For an organism, its most basic ability is to sense and interact with the environment, which is the basis to survive and explore the world.

The core foundation of autonomous handling in the construction and development of smart port is intelligent handling. The most critical technologies of intelligent handling are in two aspects: one is intelligent control of handling and transportation facilities, and the other is intelligent management of organization and implementation of operation processes. Intelligent control is based on intelligent perception and intelligent decision-making, the facility can independently identify and determine the handling objects and operation targets, and complete handling tasks safely, efficiently and automatically. Intelligent management is to achieve group intelligent control and management of a series of intelligent operation and control activities and behaviors, so as to ensure that the whole handling activities and processes are reasonable, efficient, safe and automatically completed. Intelligent control mainly solves the problems of correct control and execution of operation instructions of handling facilities, while intelligent management mainly solves the problems of correct control and execution of planning and scheduling of terminal handling system. The synergistic relationship between the two is shown in Fig. 2.2.

At present, technologies applied in most automated container terminals and automated bulk cargo terminals in the world still belong to the category of automatic control technology, and autonomous control is barely adopted, that is, the automatic control can only be implemented according to the given control model and control strategy with low intellectualized degree, while the actual process of handling activities is very complex, and has the accumulation effect of practical experience and

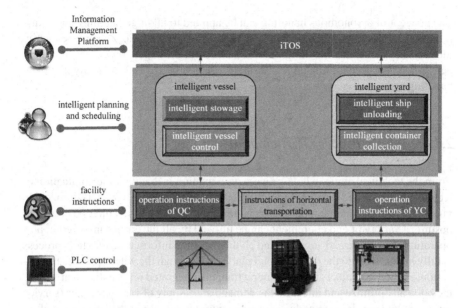

Fig. 2.2 Synergistic relationship between intelligent control and intelligent management

knowledge. Taking the handling process of a quay crane for example, an excellent QC driver can appropriately control the container movement velocity and acceleration, accurately select braking time point and braking duration, according to the visuality, somatic sensation, hand feeling, perception, experience, instantaneous judgment, to ensure that the container is operated in the trajectory matching with the terrain in the process of handling movement with fast, correct, smooth lifting and landing. The container automatic control system can only obtain physical parameters according to the configured sensor group and the given detection logic, and the control computer can calculate the output control data according to the perceived external information, the control model and decision rules set by the control system, so as to realize the automatic handling operation of the QC. Obviously, the QC automatic control system can well follow the logical rhythm in the handling process, but its control adaptability and robustness are not enough, and the intelligent degree cannot reach the level of an experienced QC driver. The autonomous handling control system combines the advantages of the two control modes and highlights the characteristics of intelligent control.

Intelligent perception and intelligent control have given autonomous handling system the elements of artificial intelligence and abilities. Autonomous handling has guaranteed the flexible connection of the system to the outside world, as well as an important means for the automatic QC to acquire external knowledge. Autonomous handling system in the future is not just an "expert" of automatic handling problem, also an "ordinary one" having learning ability to deal with all of the handling systems.

The concept of autonomous handling can be applied to all large handling and transportation facilities in smart port, so that these mechanical facilities can independently identify and determine handling objects and operation targets based on intelligent perception and decision-making, and safely, efficiently and automatically complete handling tasks.

2.2.4 Whole-Process Participation

From the technical angle, whole-process participation is that with the application of technologies such as 5G technology, cloud computing, mobile Internet, IOT, machine vision, real-time dynamic monitoring to vibration, shock, temperature, position and posture, a variety of end equipment can be utilized by all the parties involved in port operations anywhere and anytime, and all the real-time information of whole process is fully integrated into the unified cloud platform. Through the whole-process participation, extensive contact and deep interaction, the integrated information platform of the port can optimize and integrate the demand and supply of various parties (service providers and service demanders) to the maximum extent, so that the needs of all parties can get immediate response.

The whole-process participation also reflects the service consciousness, service quality and service ability of the port enterprise, which is a kind of soft environment and embodies the internal culture of the enterprise. A learning-oriented enterprise is bound to generate a distinctive internal competitiveness among all employees. Just like a family with a good tradition, the character and spirit melted in the blood can be reflected in everything done by each member. For example, in a port enterprise full of safety culture and atmosphere, the employees will always maintain safety awareness in the work without any fluke mind for convenience, which may lead to serious accidents. In addition, taking the service of an enterprise as an example, the service quality of a port enterprise does not only depend on the hardware condition, but also the service consciousness and service quality of all the staff, as well as their continuous innovative design, improvement and optimization of the service process. Every aspect of the development of smart port needs the whole-process participation of each member.

Now, the whole-process participation in the construction and development of smart port is particularly reflected in the close connection and integration of high and new technologies with port functions. Taking the intelligent supervision of the dangerous goods containers in import and export shipping as an example, the whole operation process of dangerous goods container from packing to loading involves 16 documents through 16 steps, including: entrusting order (1); packing performance inspection result sheet of inbound and outbound cargos (2); packing use inspection result sheet for dangerous goods transportation (3); technical description of dangerous goods in packaged form (4); MSDS (material safety data sheet) (5); copy of dock receipt (6); declaration on safety and fitness of dangerous goods (7); container packing certificate (8); packing list (9); authorization letter for customs

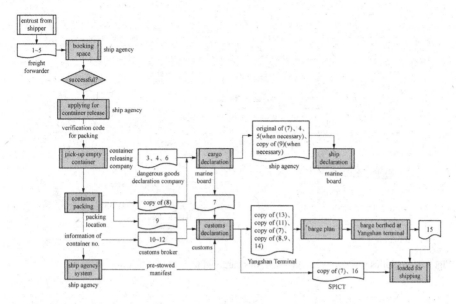

Fig. 2.3 Operation processes of export dangerous goods

declaration (10); declaration for export cargo (11); cargo description (12); shipping order released by customs (13); port dangerous goods operation declaration form (14); customs release information (15); electronic equipment interchange receipt (16). The specific operation processes are shown in Fig. 2.3.

The corresponding steps experienced in physical logistics include: dangerous goods packing in container, dangerous goods container transporting to the storage yard, storage of dangerous goods container, dangerous goods transporting to the terminal, dangerous goods container loading in vessel. The specific processes are shown in Fig. 2.4.

In this example of the intelligent packing supervision of dangerous goods container, machine vision has tested the fitness of dangerous goods for packing, physical sensing system has monitored the safety in transportation and storage process, the electronic documents have been verified for information during transformation of logistics process, etc., which fully shows the significance and value of the whole-process participation in ecological environment of smart port.

2.2.5 Continuous Innovation

Innovation is unique cognitive ability and practice ability of human. It is a kind of behavior to improve or create new things, methods, paths and environments in order to meet the needs of social and economic development and achieve certain beneficial effects by referring to the existing social and economic targets or goals, taking

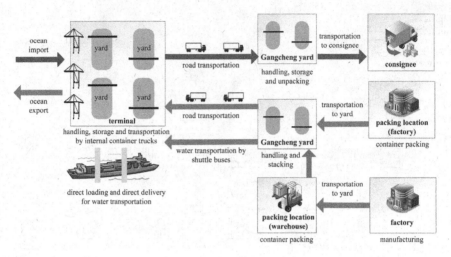

Fig. 2.4 Logistics process of export dangerous cargo with maritime shipping

the views different from the conventional thoughts as the guidance, and using the existing knowledge and materials. Continuous innovation behavior in the ecological environment of smart port can be divided into two categories. One is public innovation, which reflects that the enterprise employees and their managers, apply the accumulated experience and knowledge to discover the actual and potential needs of the working targets based on their own working environment, and promote technological innovation, service innovation and process innovation through a variety of innovative technologies and products. The other is system innovation, which is mainly reflected in the innovation of organizational management technology. It is the dynamic change of various system elements, relationship between the elements, system structure, system process and relationship between system and environment, so as to promote the continuous improvement, optimization and upgrading of the overall functions of the system.

The most important aspect of public innovation in the construction and development of smart ports is user innovation, since the user knows best what he needs to develop, what he needs to serve, what he needs to improve, what experience he has accumulated and what knowledge he has mastered. User innovation refers to taking user as the center, considering the changes of user's application environment, and mining the needs through the interaction between researcher and user. The user participates in the whole process from idea proposal to technology research, development and verification. It brings valuable innovative applications to the user through his experience and other ways, and such innovative practice also promotes the technological progress of the enterprise. Taking the innovation and development of intelligent stowage system of container terminal as example, the stowage plan is the most important and most complex operation plan of container terminal, and the stowage plan has greatly determined the quality of container stowage in vessel, the length of vessel berthing time, the service quality of container terminal, handling

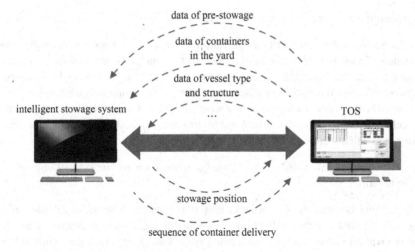

Fig. 2.5 The architecture of intelligent stowage

efficiency of container terminal, etc. So far, most stowage plans are still done manually in the TOS system. The quality of the stowage plan completely depends on the experience, knowledge, insight, logical reasoning ability and working condition of the stowage planner. In fact, when he is making the stowage plan, there is a logical reasoning and decision-making system of the stowage from his own experience, knowledge and rules hidden in his mind. With artificial intelligence technology, a stowage logic and decision system has been established parallel to the human mind, which brings together various stowage experience and knowledge for different types, different sizes, different routes and different vessels, and can automatically simulate and complete the stowage plan well with multi-subject participation and multi-factor interaction. The research and development process of intelligent stowage system is a successful practice of user innovation. Figure 2.5 shows the architecture of intelligent stowage.

Another important aspect of continuous innovation of smart port is system innovation, which is mainly reflected in the development and guarantee of innovation system. It mainly includes the following five aspects.

1. Management innovation and system innovation are the guarantee of enterprise innovation.

Management innovation is the innovation activity and process that realizes the enterprise goal through the newer and more effective resource integration. Framework innovation is to change the original enterprise framework and establish a new enterprise framework which is characterized by clear property rights, clear functions and responsibilities, separation of government and enterprise, scientific management to meet the requirements of market economic system and socialized mass production.

2.   Idea innovation and talent innovation are the foundation of enterprise innovation.

Idea innovation is the premise of all innovation activities. Idea innovation is to change and renew the ideas, which can better adapt to the change of internal and external environment of the enterprise, make more effective use of all kinds of resources, obtain profits more beneficially and seek the enterprise's further development. Talent innovation is a process in which the enterprise introduces various senior management and technical talents urgently needed and improves the overall quality of employees to form new talents through a variety of effective ways.

3.   Technological innovation and knowledge innovation are the keys of enterprise innovation.

Technological innovation is a process that starts with a new technical idea and ends with the first commercialization of new technological achievements. Knowledge innovation is the process of acquiring basic knowledge, technical knowledge and applied knowledge through basic research, applied research and development research.

4.   Product innovation and brand innovation are the carriers of enterprise innovation.

Product innovation is an activity in which the enterprise introduces products with new functions, new structures and new appearance to the market in order to better meet the needs of customers. Brand innovation is an activity in which the enterprise launches a new brand to the market in order to further improve its commercial competitiveness, shape and enhance the value of brand image, and improve brand awareness, reputation and recognition.

5.   Market innovation and marketing innovation are the realization of enterprise innovation.

Market innovation is a series of innovative activities that the enterprise opens up new market by realizing the commercialization and marketization of various new market elements. Marketing innovation is a series of innovative activities carried out by the enterprise in order to achieve business goals and satisfy market demand by improving the marketing activities.

Under the condition of construction and development of smart port, the innovative ecological environment beneficial to innovation will form, and innovation democratization will gradually become normal. Through the construction of user-centered open and collaborative innovation platform, the innovation ecology of smart port conducive to the emergence of innovation is formed by deep complementation and interaction between technological progress and application innovation system design. In short, continuous innovation of port is to enable the function of continuous innovation and self-improvement of port through extensive participation and deep interaction of parties involved in port operation, human–computer interaction between port managers and intelligent information systems, and autonomous learning of intelligent information systems. Continuous innovation is also one of the most important ecological features of smart port.

Chapter 3
Smart Port and Cyber-Physical System

3.1 General Introduction to Cyber-Physical System

Cyber-Physical System (CPS) is a new type of intelligent control system that integrates modern network communication technology into traditional digital control system and deeply combines physical system with information system through network communication technology. At present, the related technologies and applications of CPS have been not only the hot research issues in academic circles, but also promoted to the strategic level by many countries and regions and become the key research field. As a simple application of CPS, the IOT covers many fields such as medical and health care, smart home, transportation, logistics, industrial process, national grid, etc., and achieves informatization and intelligence in all areas.

3.1.1 The Concept of the Internet of Things (IOT)

CPS is a complex system that connects physical devices to the Internet and enables them to have functions of computing, communication, precise control, remote coordination and self-management through computing, communication and control (3C) technology, to achieve the real-time perception, information service, dynamic control of large system and realize the integration of virtual network and real physical world. It is an intelligent system of the new generation, integrating computing, communication and control.

CPS nowadays has spread to applications from ordinary household objects to sophisticated industrial tools, or other national and even world-class applications such as intelligent transportation, intelligent grid, etc., and in the process of implementation of these industrial applications many related independent intelligent devices with features of computing, communication, control, collaboration and autonomous management have been derived.

© Shanghai Scientific and Technical Publishers and Springer Nature   27
Singapore Pte Ltd. 2022
W. Mi and Y. Liu, *Smart Ports*, https://doi.org/10.1007/978-981-16-9889-7_3

The essence of "Industry 4.0" in Germany is to "build an intelligent factory with cyber-physical system". Zhou Ji, the President of the Chinese Academy of Engineering, put forward the concept of Human-Centered Cyber-Physical System (HCPS) in his report, *Insight into the Development Strategy of China's Intelligent Manufacturing*. He pointed out that "the traditional manufacturing process will transform from the binary system of 'human-physical system' to the ternary system of 'human-information-physical system' by means of intelligent manufacturing strategy", which profoundly revealed the connotation of the CPS. Moreover, the application of CPS in the industrial field forms the Industrial Cyber-Physical System (ICPS), and the "Industrial Internet" proposed by the United States also has the same essence.

Because of the calculation nature of CPS, devices accessed to CPS must have a strong ability to calculate. From this point, if the system architecture of CPS is compared as a fat-client-server mode, the IOT can be regarded as the thin-client-server mode, which is a simple application of CPS and has similar technical features of CPS.

The IOT means things are connected to the Internet. The International Telecommunication Union (ITU) has a clear and well-established definition for it. "The IOT can be viewed as a global infrastructure for the information society, enabling advanced services by interconnecting (physical and virtual) things based on existing and evolving interoperable information and communication technologies (ICT)." It is a network that connects things to the Internet through information sensing devices, such as QR code readers, radio frequency identification (RFID) devices, infrared sensors, global positioning systems and laser scanners to exchange information and communicate according to the agreed protocol, so as to realize intelligent identification, positioning, tracking, monitoring and management. It can be seen that the IOT has the same meaning as the CPS.

Based on the Internet, telecommunication network, private LAN and other networks, the IOT can connect all the physical objects that can be independently addressed to form a huge network, so that people, computers and physical devices can exchange information and interconnect with each other at any time, any place. The IOT is a network formed by the extension of Internet technology to the traditional physical system, and it is the product of the combination of physical system and network technology.

3.1.2 Composition and Technical Features of IOT

1. Technical architecture of IOT

According to the definition, the IOT system obtains all kinds of sensor data through sensing devices to exchange information through the network, and calculates and processes the acquired data to achieve intelligent control, independent decision-making, autonomous management and other functions. Therefore, it is generally agreed that the IOT system has a three-layered structure: the perception layer, the

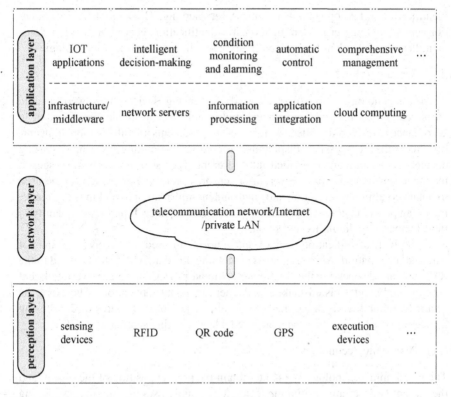

Fig. 3.1 The three-layered technical architecture of IOT

network layer and the application layer. The technical architecture model is shown in Fig. 3.1.

(1)   Perception layer

The first layer of the IOT system is the perception layer, which realizes perception, recognition and positioning of the object and its surrounding environment information through various sensor devices, RFID devices, QR code readers, infrared equipment, GPS and so on. The perception layer is the foundation of the IOT system and the source of all kinds of information. The technologies involved at this layer mainly include automatic identification technology, sensing technology, positioning technology, etc.

(a)   Automatic identification technology

Automatic identification technology, based on computer technology and communication technology, mainly realizes the functions of data acquisition, coding, identification, data management, data transmission and data analysis. At present, automatic identification technology mainly includes laser scanning, RFID, biometric identification (including fingerprint identification, iris identification, gene identification, voice identification, face identification, etc.), text identification and other technologies.

(b)   Sensing technology

Sensing technology is to utilize all kinds of sensing devices to convert specific measured signals into voltage, current, pulse and other available signal output in accordance with certain rules to meet the requirements of information transmitting, recording, processing, displaying. According to working principles, sensing devices in general can be divided into different types such as mechanical sensor, magnetic sensor, capacitive sensor, laser sensor, vision sensors, which can be used to obtain condition information and environment information of all kinds of objects, including pressure, gravity, velocity, displacement, distance, liquid level, quantity of flow, temperature, humidity and so on.

In the IOT, identification devices and sensors are used to perceive all kinds of data and information. After simple preprocessing, the data will be transmitted to the IOT terminal devices for the final computational processing to achieve intelligent control, intelligent decision-making and other functional applications. Therefore, the accuracy, reliability, dynamic tracking capability and other features of the sensing devices are the key issues to be concerned when selecting sensors in the IOT.

(c)   Positioning technology

The positioning technology is used to obtain the precise position of the object, and the current mainstream positioning technology mainly uses the satellite positioning system, including the GPS system in the U. S., the Beidou Satellite Navigation System in China, etc. Moreover, base station positioning is also an option, which is the positioning technology adopted by cellular networks in mobile communication, such as widely used 4G networks and rapidly developing 5G networks.

(2)   Network layer

The information transmission and interaction depend on the network, so the network layer of the IOT is above the perception layer, and the data from the acquisition equipment are uploaded to the network service platform, and transmitted and interacted among various nodes in the IOT.

At present, there are mainly two communication modes widely used at the network layer, namely wired communication and wireless communication. For example, wired local area network (LAN) is mostly used in the fields of smart home, smart building and smart factory, as well as in the integration of industrialization and informatization business. While the 4G network, 5G network in mobile communications, as well as the Wi-Fi in families or enterprises and institutions are all wireless networks.

The network communication technologies involved at the network layer mainly include:

(a)   Broadband network technology, including LAN technology, wide area network (WAN) technology, wireless broadband network technology, etc.
(b)   Short-range wireless communication technology, including Bluetooth technology, ZigBee technology, infrared technology, NFC short-range wireless communication technology, etc.
(c)   Long-range mobile communication technology, including cellular mobile communication, satellite mobile communication, etc.
(d)   Machine to Machine (M2M) technology, the use of which can realize the information interaction between man–machine, machine-machine, machine-system, etc.
(e)   Short-range wired communication technology, including computer communication technology, industrial communication technology, etc.

(3)   Application layer

The application layer of the IOT is ultimately attributed to the realization of various functional applications in various fields. At this layer, data obtained from the perception layer are processed, calculated and mined, and the results are applied to realize intelligent control, accurate management and independent decision-making of the physical system.

The application layer is divided into application infrastructure (or middleware) and the IOT.

(a)   Application infrastructure (or middleware)

Application infrastructure (or middleware) mostly refers to the unified package of many common capabilities into independent system software or service programs, which can be provided to IOT applications use, and the IOT users don't need extra development for these service programs (such as the drivers of external devices, supporting programs of communication protocol, etc.). Therefore, application infrastructure (or middleware) provides universal basic service facilities or supports for the IOT applications to realize functions of data storage, processing and computing, as well as interfaces for resource allocation. It is the technical foundation for the realization of IOT applications in various fields.

At present, the middleware of IOT mainly includes EPC, OPC, WSN, OSGi, CEP, etc. In addition, embedded middleware, digital TV middleware, general middleware and M2M IOT middleware are also applied. The IOT infrastructure also includes data processing and computing technologies that ultimately realize the IOT applications, such as cloud computing and big data mining technologies.

(b)   IOT applications

IOT applications at the application layer are all kinds of applications directly used by users, ranging from mobile payment, smart home, to intelligent agriculture, intelligent industry and other IOT applications in various fields.

2.   Technical features of IOT

The core of IOT is the information interaction between things and people. It is the process, in which the information of object and the environment is obtained with information identification device, then transmitted between nodes within the network, and processed at the nodes to support decision-making. Therefore, the basic technical features of IOT include three aspects: overall perception and identification, fast and real-time transmission, comprehensive application.

(1)   Overall perception and identification

According to the definition of the IOT, the perception ability of the IOT is to obtain various information of objects and their environment in the network with sensing devices such as RFID device, QR code readers and various sensors, and then express the condition or information of the perceived of things in a specific and appropriate way. The IOT relies on various sensing devices and identification devices in the network to acquire the condition information, location information and environmental information of objects in real time with a certain sampling period. The information is constantly updated with the condition of objects, and ultimately a large amount of data can be captured.

(2)   Fast and real-time transmission

The IOT is capable of transmitting the object information acquired to each node in the network in real time and accurately via various communication means such as wireless network, Internet or telecommunication network, so as to realize the exchange and sharing of information.

(3)   Comprehensive application

Comprehensive application includes sharing and exchanging resources, analyzing and processing massive data and information, and making intelligent decisions on various proposals. Not only does the IOT provide connection to sensors, it also has intelligent processing capabilities of its own, enabling intelligent control of objects. The IOT combines sensors with intelligent processing, and uses various technologies such as cloud computing, mode identification and big data processing to analyze and process meaningful data from the massive information obtained by sensors, so as to meet the different needs of different users.

3.1.3 Current Application Fields of IOT

The rapid development of IOT technology provides a new power for the development of social economy, it currently has been widely used in transportation, energy, health care, home, industry, agriculture, logistics and other fields, and formed smart home system, smart building system, intelligent transportation system, smart medical system, intelligent agriculture system, and even smart grid system of national strategic level, etc.

Industry is an important field for the application of IOT. The "Industrial Internet" proposed by the U. S. is the embodiment of the IOT technology in the industrial field. Intelligent industry based on the IOT is to integrate all kinds of sensing devices, wireless communication and network, computer technology and cloud computing, automatic control and other technologies into all fields of industrial production, so as to achieve the purpose of improving manufacturing efficiency, product quality and reducing production costs. At present, the application of IOT in industry mainly includes manufacturing supply chain management, digitalization of production process, remote online monitoring of products and facilities, environmental monitoring and energy management, industrial safety production management and other fields.

Smart logistics is the result of the IOT application in logistics industry. Based on IOT technology, with the aid of RFID technology, identification and tracking technology, satellite positioning technology, and other sensor network, information of each node of logistics can be monitored, including transportation, inventory inbound and outbound, packing, handling document flow and so on. The information of each node in the logistics process is circulated within the relevant departments through network, which realizes information sharing and collaboration between suppliers, wholesalers, retailers, as well as integration of logistics transportation process, storage, packaging, handling and intelligent scheduling and management of logistics process. By integrating the core business processes of logistics, smart logistics strengthens the rationalization of logistics management, reduces logistics consumption, and achieves the purpose of reducing logistics costs and increasing profits. Based on the technology of the IOT, the smart logistics system can realize the intelligent integration of the logistics system, procurement system and sales system of the enterprise, and can further integrate smart logistics with intelligent production and intelligent supply chain. The enterprise logistics is integrated into the enterprise operation and becomes an important step to build the intelligent enterprise. At present, the typical applications of smart logistics include intelligent warehouse management, intelligent freight transportation system, and personalized analysis of logistics enterprises.

Supported by the IOT technology, 5G and mobile Internet technology, cloud computing, big data technology, artificial intelligence, system simulation and preview technology, virtual reality (VR) and augmented reality (AR), handling machine vision and autonomous control and other high-end technology, smart port can realize a comprehensive perception of the terminal and data integration management. Based on perceptual collection of basic decision information, decision objectives and constraint conditions are determined, and decisions on overall planning, complex planning, dynamic scheduling and other problems can be made effectively and efficiently. On the basis of intelligent decision-making, the facility can identify and determine the handling targets and operation objectives independently, and complete the operation tasks safely, efficiently and automatically. With the application of cloud computing and mobile Internet technology, relevant departments of the port can utilize a variety of terminal devices anytime and anywhere to fully integrate into the unified cloud platform. Through extensive contact and interaction, the comprehensive information platform of port can optimize and integrate the demand and supply of various parties to the maximum extent, so that the demands of all parties can get immediate response. The port is equipped with the function of continuous innovation and self-improvement via the participation and interaction of related parties, the human–computer interaction between port managers and intelligent information system, and the autonomous learning of intelligent information system.

The functions of smart port include the autonomous operation system of terminal handling facilities, intelligent management system integrating facility operation and operation planning, intelligent government business system between port and customs, inspection and quarantine, revenue and other departments, intelligent business system for exchanges between port and railways, highways, agency and logistics parks, etc.

3.2 Development of Cyber-Physical System

The term "IOT" was first put forward by Professor Kevin Ashton, Massachusetts Institute of Technology, U.S. in 1991, and Bill Gates also mentioned the IOT in his book *The Road Ahead* in 1995. The Academy of Sciences in China initiated the study of "Sensor Networks" in 1999, and proposed that "sensor network is another development opportunity for mankind in the next century" in the International Conference on Mobile Computing and Networking held in the U.S. In 2009, the European Commission published the European IOT action plan, which predicted the application prospects of the IOT technology and proposed that the EU should strengthen the management of the IOT in order to promote its development.

In China, the development and research of the IOT have also received great attention. In August 2009, Wen Jiabao, the former premier of the State Council, put forward the concept of "perceiving China", which promoted the research of the technology and application of the IOT in China to a climax. Until then, the IOT in China as a national strategy continued to develop, the Ministry of Industry and Information Technology in 2011 and 2012 consecutively issued a scale of 500 million yuan of special funds, and in 2012 the National Development and Reform Commission also launched 600 million yuan of special funds for the research, development and industrialization of IOT technology. In 2013, the State Council issued the *Guidelines of the State Council on the orderly and healthy development of the Internet of Things*, to guide the development of IOT from the level of overall national planning. More than 30 provinces and cities across China have taken the IOT as a priority for the development of emerging industries in the region. They have issued many special plans or action themes, and carried out the technological research and application development of the IOT in various fields such as industry, agriculture, home appliances and medical care.

At present, the application of the IOT has been all over the military, power, industry, agriculture, construction, medical care, environmental monitoring, space and ocean exploration and other fields, especially the deep research and development of artificial intelligence has promoted the IOT technology to a large system of CPS.

3.3 Applications of Cyber-Physical System in Smart Port

The development of CPS and IOT technology has been widely applied in many fields, including transportation, logistics, ports and other fields, providing solid technical support for the construction and development of smart port. At present, a variety of independent application systems with the ultimate goal of building smart port system have emerged.

3.3.1 Whole-Course Tracking and Ship-Shore Docking of Dangerous Goods

In recent years, the type and quantity of dangerous goods transported by water are increasing, which brings the corresponding increase of risks. For the dangerous goods on board, burning, explosion and other malignant accidents of dangerous goods containers often occur. In order to effectively avoid the occurrence of accidents, in addition to fulfilling its supervisory function, it is necessary for the Maritime Safety Administration to make a systematic analysis and summary of the foregoing steps of dangerous goods on board, and realize the whole-course tracking and ship-shore docking of dangerous goods.

The whole-course tracking and ship-shore docking of dangerous goods take the IOT technology as the core, which is realized by integrating the IOT technology with 5G technology and image identification technology. The specific implementation is to install monitoring cameras inside and outside the vehicles transporting dangerous goods containers, which can enable the regulatory authorities to monitor the dangerous goods containers transportation process in real time, understand the driving condition and operating environment of the dangerous goods vehicles, and discover environmental or human problems in time. With the help of mode identification technology, speeding, phone calls when driving and other illegal behaviors, as well as dozing, inattention and other fatigue phenomena of drivers can automatically be identified and alarmed. Meanwhile, 5G technology and IOT technology are used to monitor the condition information of dangerous goods containers in real time through hardware such as on-board data acquisition sensors, so as to realize the round-the-clock supervision of people, vehicles and goods.

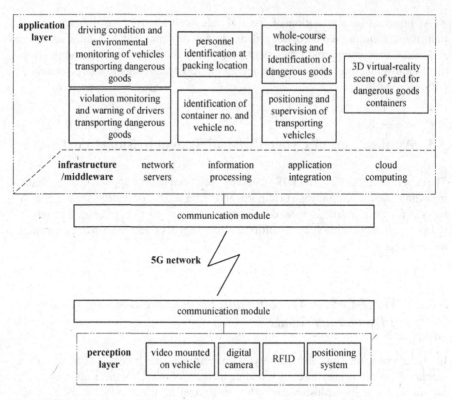

Fig. 3.2 The architecture of the whole-course tracking and ship-shore docking system for dangerous goods

The architecture of the whole-course tracking and ship-shore docking system for dangerous goods is shown in Fig. 3.2.

The system has adopted complete set of whole-course tracking regulation technology for dangerous goods containers, including 5G whole-course tracking regulation technology, intelligent identification and wireless sensing technology, positioning-monitoring technology and virtual reality technology, constructed efficient and practical whole-course supervision platform for the dangerous goods, improved the application level of intelligent perception and active warning, realized organic integration and sharing of the various regulatory supervision information of various regulatory departments to enhance the regulation ability and solve the blind spot in the tracking and supervision of dangerous goods on board at present.

1. Whole-course tracking regulation technology

The traditional way to realize the dynamic monitoring of high-risk vehicles in China is to make use of the built-in timing function of the terminal on-board. Generally, the continuous driving time exceeding 4 h is seen as fatigue driving. However, this method cannot effectively monitor multiple drivers assigned in the same vehicle, and ignores the individual differences of driver fatigue, resulting in the lag of fatigue driving supervision.

In the future, the transmission rate of 5G network can reach up to 10 Gbps, which will effectively solve the problems of slow response and poor monitoring effect of existing video surveillance, and can swiftly provide monitoring data with higher resolution ratio. After installing the monitoring camera inside and outside the vehicles transporting dangerous goods containers, the supervision department can monitor the transportation process of dangerous goods containers in real time, understand the driving condition and operating environment of the dangerous goods vehicle, and discover environmental or human problems in time. With the help of mode identification technology, speeding, phone calls when driving and other illegal behaviors, as well as dozing, inattention and other fatigue phenomena of drivers can automatically be identified and alarmed. Meanwhile, 5G technology and IOT technology are used to monitor the condition information such as location, temperature, pressure, concentration of dangerous containers in real time through hardware such as on-board data acquisition sensors, so as to realize the round-the-clock supervision of people, vehicles and goods.

2.   Intelligent identification and wireless sensing technology

The application of intelligent identification and wireless sensing technology mainly includes personnel identification at packing location, identification of container no. and vehicle no., whole-course tracking and identification of dangerous goods, positioning and supervision of transporting vehicles, and the whole-course visual supervision of dangerous goods containers.

(1)   Personnel identification at packing location

Face identification technology is mainly adopted in the personnel identification at packing location, to verify the identity and qualification of the supervisor, and standardize the operation process of the supervisor. Face identification technology will be fully enabled at packing location, and facial images and corresponding professional qualifications of declaration clerks and packing supervisors will be collected in advance, and files will be established. When packing, the packing supervisor enters the packing location by "face", and the identity and professional qualification of the packing supervisor are confirmed through face identification. This technical "face" identification can automatically verify the identity and qualification of the supervisor in a short time, preventing the situation of packing without supervision or the operation by the ordinary personnel without supervision qualification, so as to avoid the hidden danger caused by the non-standard packing process in the container packing.

(2)   Identification of container no. and vehicle no.

Identification of container no. and vehicle no. is the digital integration of advanced computer mode identification technology, image acquisition technology, network technology, etc. When container transportation vehicles passing the identification area, without human intervention, the information of vehicle license plate (license no., license plate color, plate type, etc.) and container information (container no., container size, container door, etc.) can be automatically identified via container video image and vehicle video image acquired by the high-definition cameras as well as image data processed by the background pretreatment system.

Based on vehicle video images provided by high-definition cameras in virtual toll station, as well as intelligent monitoring and recording systems of vehicles on road, supervision network for vehicles transporting dangerous goods containers covering provinces and cities has been built, with handling locations as "points" and transporting routes as "lines" is constructed, which can effectively identify the license plate no., container no. and other regulatory information of vehicles transporting dangerous goods containers on the highway; automatically verify the customs supervision information and identify risks; complete management operations of intelligent monitoring and recording systems, such as automatic verification of vehicle tracks, cargo inspection or vehicle release instructions. It can realize the diversion and automatic inspection and release as well as real-time whole-course management and control for the vehicles transporting dangerous goods containers, and comprehensively improve the safety supervision level of vehicles transporting dangerous goods containers.

(3)   Whole-course tracking and identification of dangerous goods

RFID technology can be used to track and monitor the route and transportation time of vehicles. RFID tags have been stuck on the packages of dangerous goods before transportation. When passing the gate, the RFID reader will automatically collect the information of the goods, check the amount of the goods and input the information into the database of the host system. After storage, the reader will automatically complete the inventory counting operation and update the inventory information. When the goods are loaded into container, the RFID reader can automatically update the inventory information and track the goods in real time.

During the transportation of dangerous goods, with the help of RFID monitoring points set along the way, the location and completeness of the goods can be accurately understood. When there is an accident, if the driver does not know his position, the RFID monitoring points can also quickly locate the accident, so as to ensure that the emergency rescue measures can be taken in time.

(4)   Positioning and supervision of transporting vehicles

Positioning equipment must be installed in the vehicles when transporting dangerous goods. With the aid of mobile communication technology and GPS positioning technology, as long as they can be covered by satellite signal, real-time and whole-course tracking, monitoring and management of the position, velocity, orientation, route, travel area and other condition information of vehicles transporting dangerous goods containers can be effectively realized.

(5)   Whole-course visual supervision of dangerous goods containers

With the help of 3D visualization technology, a digital monitoring platform for the whole-course tracking of dangerous goods containers is established to realize holography and visualization of the whole-course supervision of dangerous goods containers.

The management system of yard for dangerous goods containers based on 3D virtual-reality technology, can establish digital entity models of facilities in proportion, such as blocks, office facilities, fire-fighting facilities, entrance and exit gates, sewage and other important facilities, according to statistical information of existing facilities, to ensure that the virtual scene of the yard and the actual scene can match and the basic data are real and reliable, providing for operators a fundamental platform to supervise the yard for dangerous goods containers comprehensively with multiple perspectives in a 3D way.

The system also configures the fixed fire-fighting facilities and equipment in the yard according to the real location and properties to realize the functions of query and visualized management of fire-fighting facilities and equipment in the platform. Administrative staff can not only arbitrarily query the location and properties of fire hydrants, fire guns, fire engines and other facilities, but also dynamically demonstrate the effective radius of fire equipment through 3D scenes to assist fire personnel to conduct fire emulation and improve the emergency rescue response ability of fire personnel. In addition, through the hidden operation to the model on the ground, it can also display the underground sewer pipeline layout and related parameters of the yard, so as to assist the operators in the yard to make the correct judgment and decision to avoid the situation of untimely or improper disposal.

3.3.2 Identification Analysis of Industrial Internet in International Multimodal Transportation of Containers

Due to large number of participants, low industrial concentration, restricted cooperation in the industrial chain of international multimodal transportation of containers, many problems have been accumulated in the process of its development, including non-uniqueness of container identification, manually input container no., serious information asymmetry between suppliers and demanders, etc., which have greatly limited the healthy development of the industry.

It is an important way to solve the problems mentioned above by establishing the second-level identification analysis nodes of industrial Internet for the international multimodal transportation of containers, and providing identification analysis services to the key industries in the whole container supply chain, such as container manufacturing, shippers, ports, shipping companies, logistics services, finance and insurance, etc.

As the application of IOT in the industrial field, the industrial Internet can be linked up to information platform and business platform via unique identification and data collection in multimodal transportation logistics, to achieve real time and transparent information of the whole supply chain in the global environment, and remote monitoring, tracking and management of the multimodally transported containers in the whole supply chain.

In addition to basic capabilities, such as identification registration, identification analysis, identification agent service, data synchronization and so on, the second-level identification analysis nodes of industrial Internet in international multimodal transportation of containers can also be applied beyond the basic capabilities, including applications in supply chain management, the whole-life-cycle management, the whole-course traceability, management of finished product containers, cross-border logistics of multimodally transported containers, forwarder transportation management of railway containers, supply chain management of dangerous chemical containers, container maintenance management and container operation management of yard in port terminals, etc.

In the architecture of the second-level identification analysis nodes of industrial Internet for the international multimodal transportation of containers, the application layer, including infrastructure layer, middleware layer and capability application, is emphasized.

The infrastructure layer provides services such as network servers, databases, cloud computing and virtualization. The middleware layer includes communication interface, standard registration management, identification analysis service and operation security maintenance, etc. It mainly realizes the data exchange between the second-level nodes and the first-level (national-level) nodes.

Capability application includes basic capability application and extended capability application.

The application of basic capability is embodied on the second-level-node platform, which mainly includes core software and hardware systems such as identification registration, identification analysis, data synchronization, business management capabilities and security assurance. Built on Tianyi Cloud, the second-order platform is a flexible and extendable platform of industrial Internet and big data, which can provide container manufacturing enterprises with a series of services such as data collection, equipment monitoring, data storage and analysis, operation optimization and resource management.

The application of extended capability is embodied on three platforms: identification convergence and enabling platform, industrial application platform of second-level nodes, and IOT convergence and enabling platform. The main functions include: supply chain management, the whole-life-cycle management, the whole-course traceability, management of finished product containers, cross-border logistics of multimodally transported containers, forwarder transportation management of railway containers, supply chain management of dangerous chemical containers, container maintenance management and container operation management of yard in port terminals, etc.

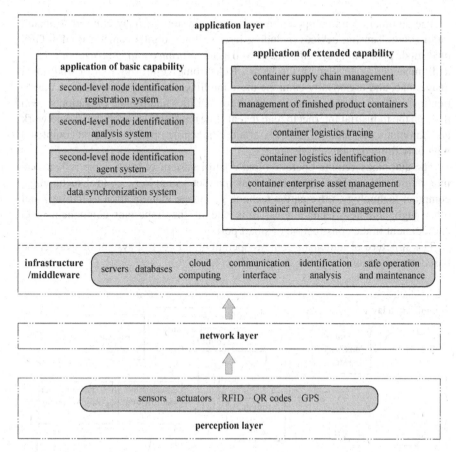

Fig. 3.3 The architecture of the second-level identification analysis nodes of industrial Internet for the international multimodal transportation of containers

The architecture of the second-level identification analysis nodes of industrial Internet for the international multimodal transportation of containers is shown in Fig. 3.3.

3.3.3 Automatic Remote Operation and Control of Quay Cranes

The operation of the traditional quay crane requires the driver to follow the operating object throughout the operation in the cab moving along with the quay crane. Due to the reasons of equipment vibration, inertia, high concentration of the driver and so on, the driver's labor intensity is high, while the efficiency is low.

The automatic remote operation and control system of quay crane is to install multi-source sensing devices including laser scanner, digital camera, Beidou-GPS dual-mode positioning system, etc., as well as monitoring equipment, network server and embedded controller on the basis of electronic control system of traditional quay crane, and organically combine these intelligent core components with operation instructions, chassis, container handling facilities and remote operation station through the industrial Internet to realize real-time data interaction within the network and automatic operation of the quay crane. The whole operation can be completed with only limited manual assistance.

The automatic remote operation and control system of quay crane reflects the integration and application of industrial Internet technology, control technology, computer technology, information technology, image identification technology and other high and new technologies. It is one of the important components and embodiment in the construction of smart port.

The structure of automatic remote operation and control system of quay crane based on industrial Internet is shown in Fig. 3.4.

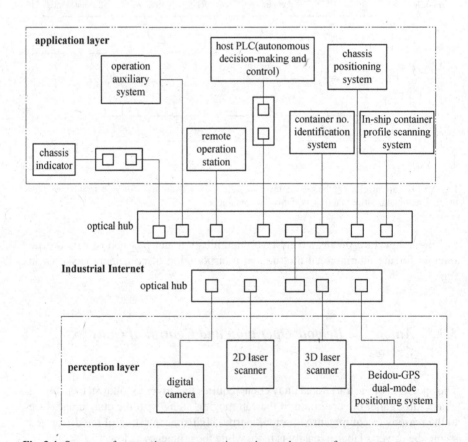

Fig. 3.4 Structure of automatic remote operation and control system of quay crane

The application layer of the remote operation and control system includes several functional subsystems, such as remote operation station system, video monitoring system, remote operation auxiliary system, chassis positioning system and scanning container profile system. Each subsystem performs its own duty and coordinates with each other to jointly complete the remote control of the quay crane.

1.   Remote operation station system

The remote operation station (ROS) is installed in the control room, including displayer, handle, touch screen, etc. The ROS is connected to the master PLC for remote control in the control room through optical cables, as well as to the control system on the quay crane. The local PLC in the quay crane control system communicates with the host PLC, receiving instructions from the host PLC and transmitting all kinds of data and information required to the host PLC. The driver can control the facility through the ROS with the help of video displayer and operation auxiliary system (OAS) to ensure correct, safe and efficient handling operations.

2.   Video monitoring system

The video monitoring system monitors the running condition of each operating mechanism of the quay crane during remote operation with a digital camera, and the facility condition will be displayed in real time on the monitoring interface of the remote-control operation auxiliary system. The monitoring includes: whether the operating chassis has arrived, the no. of the operating chassis, the personnel in the lane under the quay crane and other non-operating chassis are in the safe condition. The information of the chassis to be operated under the spreader and the lock-hole position of the lower container are also monitored, so that the remote driver can realize correct container landing and locking operation according to the relevant pictures. The monitoring also includes the operation of lifting spreader, the container landing and matching as well as the pitch hook of the girder.

3.   Remote operation auxiliary system

The OAS uses a graphical interface to display the facility condition and various monitoring data sent by the video monitoring system and other subsystems in real time, which is the factual reference for the driver to correctly control the quay crane and handle various emergency situations.

The system is connected with scanning container profile (SCP) system, chassis positioning system (CPS), container no. recognition (CNR) system and host PLC respectively by optical cables.

The information interaction between OAS and other subsystems includes: sending control instructions (lane number, starting scanning, etc.) to CPS; receiving the location information of the chassis from CPS; receiving the container contour information scanned by SCP, and calculating the bay position in the ship by the contour information; receiving running condition information of host PLC of the crane; according to different operation processes, sending PLC control instructions (such as getting ready, target address, container size, etc.); communicating with the CNR system, comparing the container no. from camera image recognition with the container no. given in the instruction from the operation management system to ensure correct operation.

4.   Chassis positioning system

The CPS uses a 3D laser scanning system to determine the exact location where the chassis should stop, and displays the exact location on the screen to instruct the chassis driver to park correctly and efficiently.

The CPS includes functional components such as a 3D laser scanner composed of a rotating platform and a 2D laser diastimeter, a positioning data processing system and a chassis indicator. The 3D laser diastimeter system scans and detects the location of the chassis in real time, and sends the data to the chassis positioning data processing system. The actual location of the chassis is calculated and transmitted to the OAS and the host PLC. The chassis indicator displays the chassis location to guide the driver to park correctly and efficiently.

5.   Scanning container profile system for ship stowage

SCP adopts 2D and 3D laser scanners installed on the trolley frame to implement sector scanning on the areas within the running range of the trolley spreader, and accurately measure the profile of objects within the moving range of the trolley spreader in real time, including the profile of objects in the running direction of the trolley and the profile of objects in the running direction of the gantry, and establish the profile map based on the trolley coordinate system. The computer for processing laser scanner data is used to calculate the distribution of the container stowage in the ship and calculate the tilting angle and guide frame height of the ship. All these real-time scanning data and processing results are transmitted to the OAS and the host PLC. On the one hand, the driver can master the profile condition of the container, on the other hand, the host PLC can execute intelligent decision-making according to the corresponding information, optimize the movement trajectory of the spreader, avoid possible collision, realize the trolley collision avoidance and automatic optimization of operation route for the purpose of accurate handling.

Meanwhile, the laser scanner is used to detect the drift of ship waiting for loading and unloading, and the information is sent to the host PLC to adjust the positioning target of the trolley and the lifting mechanism in real time.

It can be seen from the practical application cases in port that the CPS provides a new infrastructure and good information conditions for the realization and application of comprehensive perception, intelligent decision-making, intelligent management, intelligent control, intelligent service and other functions of smart port.

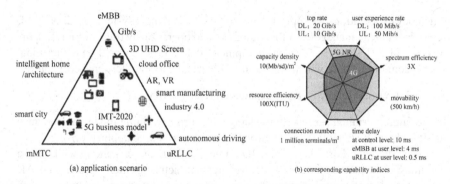

Fig. 3.5 Application of 5G communication technology

3.3.4 Application of 5G Wireless Communication Technology in Smart Port

5G communication technology is most likely to produce high-value applications in the port industry in three directions. Diversified application scenarios defined by 5G and corresponding capability indicators are shown in Fig. 3.5.

1. 5G+ intelligent driving of new AGV or Quasi-autonomous chassis

In recent years, Internet companies such as Google and Baidu have carried out tests on technology architecture for the application level of autonomous driving and Internet of vehicles communication equipment in the passenger car market in many parts of the world. Nowadays, rapid progress has been made in autonomous driving platform, vehicle-to-vehicle and vehicle-to-road coordination standards and technical equipment. The autonomous driving of heavy container chassis has also been carried out in many places around the world. At present, the problems mainly concentrate in the road-testing license of automobile companies, the standard and technical optimization of C-V2X facilities and equipment inside the chassis and on the roadside, as well as the quality of wire control of heavy chassis are still at the testing stage. Once these policies or technical difficulties are conquered, leading automakers in technology architecture of autonomous driving or Internet companies will ascend autonomous driving into the large-scale application by simplifying the experimental scenario (such as scenario data obtained from extensive tests in closed area on the open road or closed container port), along with high intensity AI training unit.

2. 5G + big data + artificial intelligence (AI)

First of all, technology-based companies with TOS can obtain mechanical operation conditions with high frequency and low delay through efficient means of 5G communication. After data transmission, if big data and AI can be effectively used and real-time dynamic scheduling can be implemented for areas of intensive operations, it will effectively help port enterprises to save energy, reduce costs and increase efficiency in operation. Secondly, other terminal management systems, such as EAM, can carry out lean management for equipment and operation by transmitting back the readings of IOT sensors of various equipment and facilities located on the terminal via 5G communication, which will further enhance the level of terminal cost control.

3.   5G + virtual reality (VR) + augmented reality (AR)

In the future, terminals are likely to change the current training mode, maintenance mode and remote operation mode through the scenario-based application of 5G. For instance, Shanghai Port has realized 4D immersive RTG driver training with VR. In the future, if the pilot experience of intelligent manufacturing enterprises in 5G + AR can be fully learnt, it will completely change the current repair mode of maintenance workers of large machines.

At present, some terminals adopt 5G high bandwidth to test massive video back-transmit of large mobile facilities in the yard, in order to realize remote control by wireless backtransmit or make redundancy for the wired network on the facility. But the bandwidth available under the current 5G spectrum is hardly sufficient for large-scale video of remote control or with very low-cost performance. It is expected on the one hand that late 5G or even 6G with enhanced standard can break through the bandwidth bottleneck, on the other hand that the development of OCR technology and 8 K camera technology will significantly reduce the amount of installed video cameras. If supplemented with edge calculation and other technical means, a large number of facilities in traditional terminal and storage yard will be able to fully realize remote control, reducing the labor intensity of drivers and the number of drivers, improving the efficiency of the terminal, and benefitting the whole industry.

Bibliography

1. Lee EA (2008) Cyber physical systems: design challenges. In: 2008 11th IEEE international symposium on object and component-oriented real-time distributed computing (ISORC). IEEE, pp 363–369
2. Lee J, Bagheri B, Kao HA (2015) A cyber-physical systems architecture for industry 4.0-based manufacturing systems. Manuf Lett 3:18–23 (2015)

Chapter 4
Smart Port and Middle-Office System

By abstracting the business, data and technology, the middle-office system can reuse the service capability and construct a unified, standardized and institutionalized digital service system. The middle office is a kind of technical architecture, but also an idea, a strategy. The digital middle office of port is the exploration and practice of the middle office strategy in the port and shipping field. The digital middle office of port reuses the digital service capacity of the port by abstracting the internal and external business, data and technology of the port, to form the port-level or group-level service capability, eliminating the barriers between various business departments, terminals, depots and other subsidiary companies in port enterprises, and adapting to the development strategy of diversifying business of large port enterprise groups. Based on the digital middle office of port, the front office applications for port operation managers or final customers can be developed quickly, so as to meet the front office needs with various personalized features, providing a clear path for the digital transformation of port enterprises, and enabling the construction of smart port.

4.1 General Introduction to Middle Office

Middle office is a new concept, but it is an old term. In the new period, we give it a new connotation. In the Eastern Han Dynasty in ancient China, Shangshu office became the center of the government, known as Zhongtai (which means middle office). The improved system of Three Councils and Six Boards in the Tang Dynasty also called Shangshu office as Zhongtai.

In the organizational structure of an investment bank, the front office is the position that directly interacts with customers, such as lobby managers, account managers, tellers, etc. The middle office refers to all staff who directly support the front office, using the resources of the front office or back office to provide professional management and guidance to the front office, as well as risk control, such as risk management, compliance management, financial management and control, IT services, etc. The back office refers to functional positions at the backstage that perform management functions, such as settlement, clearing, accounting, human resources, etc.

Supercell, a Finnish mobile game company, has organized several development teams with a small front office, and each team contains various roles needed to develop a game, which allows the team to make quick decisions and develop quickly. However, the infrastructure, game engines, internal development tools and platforms are provided by "tribes", which can be expanded into several subgroups with common goals when necessary, but do not offer games to consumers.

By 2015, Alibaba had already had a large number of individual members and enterprise members. The business types were complicated, business contents were cross-dependent, and business teams were numerous, which could not timely respond to business requirements. Therefore, in December, Daniel Zhang, the CEO of Alibaba Group, announced the launch of Alibaba's 2018 Middle Office Strategy through an internal email, to build a more innovative and flexible organizational mechanism and business mechanism of "Big Middle Office, Small Front Office" in line with the DT era, and realize the innovation of management mode. That was, splitting the product technical power and data operation capability off from the front office and establishing an independent middle office, including the search division, sharing business division, data platform division, etc., to provide services for the front office, namely, the retail e-commerce business group. In this way, the front office could be streamlined to maintain enough agility, and business development and innovation needs could be better met.

In May 2017, the book *The Way of Enterprise IT Architecture Transformation: Alibaba's Middle Office Strategic Thinking and Architecture Practice* was published, explaining in details that the middle office of the business is between the front office and back office, which uses the shared way to build, and has solved the problems of repeated development, dispersed data and high cost for trial-and-error in the previous stovepipe-type and single framework design. The book listed some of the principles of building a business middle office, including high cohesion and low coupling, data integrity, operationability, and gradualism. The publication of this book had promoted the development of middle office thought and the construction of middle office.

Since then, a lot of Internet companies quickly followed up in the middle office construction. DiDi shared *How to Build DiDi Business Middle Office* in December 2017. JD.com announced in December 2018 that it would adopt an organizational structure of front office, middle office and back office.

4.1.1 Service Mode of Middle Office

Middle office can be used as a mode and concept of enterprise organization and management. However, from the point of technical system, the middle office can also be used as a new architecture of enterprise IT infrastructure. In addition, in order to build the middle-office system, some enterprises will form a special technical team of middle-office to be in charge of realization and operation. Therefore, as an organizational management mode, the middle office and the middle-office system are not completely separated.

The organization mode of middle office is to build a unified and collaborative platform within the company. On the one hand, each business department can maintain relative independence and decentralization to ensure the sensitivity to the business and innovation; on the other hand, a powerful platform is used to implement overall coordination and support for these departments, balancing centralization and decentralization, and providing growth space for new businesses and new departments, so as to significantly reduce the cost of organizational reform. The middle office departments refine the common needs of each business line to minimize the repetition of "making wheels".

From the level of technical system, middle office is also a sharing service platform of enterprise-level. Traditional IT systems or components do not pay much attention to the reuse and sharing of system capabilities, so enterprises have introduced and built multiple sets of stovepipe-type systems with repetitive functions in the process of informatization for many years. However, middle office requires fine-grained analysis of capabilities, identification of shared capabilities, and building shared capabilities into a unified platform.

To sum up, middle office is the hub of capabilities and the sharing of capabilities. Middle office is to build decentralized and connected businesses on a centralized basis, and provide unified services for each business. Therefore, all the platforms are the middle office, which transform various resources of the enterprise into the capability to be easily used by the front office and serve for the "user-centered" digital transformation of the enterprise. However, it should be noted that the corresponding middle office team built to match this cannot be regarded as a resource-sharing team. The middle office team focuses on how to form basic services to facilitate the construction of business applications for the front office team. Therefore, the middle office should realize the separation of platform logic from business logic, and isolate different front office businesses.

In addition, the middle office is not a microservice, since the middle office is not only a technical framework, but also the overall framework reference for the digital transformation of the enterprise. From the technical point, it can be considered that microservice is the best practice of building a middle office. Microservice is to divide the monomer architecture of J2EE era into multiple technical architectures to provide microservice. Microservices bring together related business logic and data to form separate boundaries. Microservices communicate with each other through standard protocols, such as HTTP RESTful style. Microservices are loosely coupled to each other. Different microservice development teams could theoretically use different technology stacks to implement microservices without having to be consistent. In addition, the data storage required by the microservice is generally isolated by individual database instances or database schemas, and the data interaction can only be realized through interfaces or messages, and the data of another microservice cannot be directly accessed at the database layer. Microservices emphasize the isolation principle of interfaces and encapsulation via interfaces. Since microservices can be deployed separately, the required microservices can be scaled up and down according to the needs, without targeting the whole system, which makes the system more flexible in scalability and more capable to cope with large traffic concurrent scenarios, such as seckill. Microservices have inherent features of separate development, separate deployment, separate publishing, supporting high concurrency and high availability, as well as decentralized management and other advantages. However, due to the distributed programming of microservice, it increases the difficulty of development, debugging, deployment, operation and maintenance as well as the complexity of service management, and needs to redesign the principles previously guaranteed by a single database. Although microservice put forward higher requirements for the development team, it promotes the integrated operation and maintenance ability of the R&D team, thus in turn changes the organizational structure of the enterprise's R&D.

4.1.2 Technical Connotation of Middle Office

In programming, a function encapsulates a frequently used piece of code and then it can be called directly when necessary. Using functions reflects the guideline of modular programming, which is to decompose a big problem into small problems and solve the big problem by solving the small problems. Second, using functions greatly reduces the workload required to write program segments repeatedly. The associated set of common functions can be compiled into dynamically linked libraries and class libraries, again upgrading the possibility of reuse. Since we can encapsulate some reusable procedures in the form of functions and class libraries, we can also provide reusable functions as a service. The middle office is a higher level of reusable encapsulation than functions and class libraries, thus better serving the business. So, from a technical point, the platform that provides shared capabilities in the form of service is a middle office. The three layers of sharing is shown in Fig. 4.1.

Fig. 4.1 Three layers of sharing

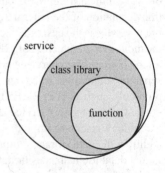

4.1.3   System and Classification of Middle Office

The middle office is generated from the sharing requirements of several similar business applications of front office, so the first proposed middle office is the business middle office. The business middle office is integratedly planned in terms of overall strategy, business support, connecting industry customers and business innovation. Therefore, the main business of the port is included in the business middle office. Business middle office is more focused on how to support the online business.

Data are generated from the business system, and the business system also needs the results of data analysis, so can the data storage and computing capacity of the business system be extracted and provided by a separate data processing platform? Thus, it not only simplifies the complexity of business systems, but also allows each system to focus on what it is good at with more appropriate technologies. This dedicated data processing platform is known as the data middle office.

Both business and data middle offices are formed in the process of architecture evolution of enterprise IT system, and the common capabilities are extracted from planning, construction, operation, maintenance and other practices of the enterprise's own IT system for many years. As two wheels, the business middle office and the data middle office build the digital middle office side by side, supporting the front office to provide customers with closed-loop services from marketing promotion, transaction transformation to intelligent service business, and promoting the improvement and development of enterprise business, as shown in Fig. 4.2.

The business middle office abstracts, packs and integrates the resource of the back office and transform them into easily reusable and shared core competence

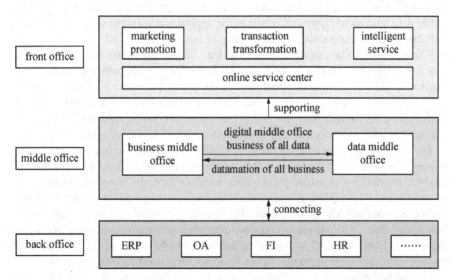

Fig. 4.2 The digital middle-office system consisting of the business middle office and the data middle office

for the front office, realizing the transformation of the back-end business resources to easily reusable capability for the front office. The shared service center of the business middle office provides unified and standard data, which are directly applicable, reducing the interaction between systems and the cost of cooperation between teams, and providing a powerful "gunfire support" capability for the front office applications.

The data middle office can access to business middle office, back office and other third-party data, complete the storage, cleaning, calculation and summary of massive data, and constitute the core data capabilities of the enterprise, providing a strong support for data-based customized innovation of the front office and data-feedback-based continuous evolution of the business middle office. It can be believed that the data middle office provides a powerful "radar monitoring" capability for the "battlefield" of front office, understanding and controlling the situation in real time. The data processing capabilities provided by the data middle office and the data analysis products built on it are not limited to serving the business middle office. The capabilities of the data middle office are open for use to all business parties. The data middle office is the support of the business middle office. The data middle office not only provides analytical functions, but more importantly provides services for the business. The data middle office often needs to process a series of data, in addition to integrating user data, orders, evaluation and other behavioral data generated by the business system, and finally provides support in the form of microservice. Meanwhile, big data processing technology is only the infrastructure provider of the data middle office. The rapid development of big data technology accelerates the maturity of the data middle office strategy and it can even be said that big data platform is the cornerstone of the construction of the data middle office. The data middle office is a platform that connects the computing and storage capacities of big data with technology and connects the application scenario capability of data with business. "Connectivity" is the essence of the data middle office.

The role of the data middle office is to lead the business, build intelligent data processing platform with standard definition that can be globally connected and extracted, and the construction goal is to efficiently meet the needs of data analysis and application for the front office. The data middle office covers multiple levels of systematic construction, such as data assets, data governance, data model, vertical data center, global data center, extraction data center, data service, etc.

From the perspective of front office application, the direct-to-user service capabilities provided by the business middle office and the prediction, monitoring and analysis capabilities provided by the data middle office are integrated, instead of independent of each other. Business middle office and data middle office complement each other and support each other. For business parties, they generate data when consuming data, and continue to consume data when generating data, thus forming a closed data loop. Business middle office and data middle office are only different in the way of technology implementation. Together, they constitute the digital middle office and becoming the two "wheels" supporting business innovation. Neither can be dispensable.

4.2 Development of Middle Office

4.2.1 Industry Developing Condition of Middle Office Technology

Let's take a look at the origin of middle office, the driving force and achievements of Alibaba Group's construction of middle office. In 2008, due to the isolation between internal departments and relatively different business objectives of Alibaba Group, Taobao.com and Tmall.com were built as two independent systems, namely two independent "stovepipe", systems. However, the basic business of them is e-commerce transaction, so the basic functions are similar, including goods, transaction, payment, evaluation, logistics, credit and forum functions. Due to the isolation between the systems, even though the flow and transactions of Tmall.com continue to fall, it is unable to lead the flow of Taobao.com to Tmall.com. Therefore, the two business departments discussed to connect the two e-commerce platforms, so as to set up a shared business division, starting the internal project known as the "multicolored stone". The results of the "multicolored stone" project, namely the shared business service center now known as the "middle office", laid a solid foundation for the rapid development of Tmall.com in the future. The middle office integrated Alibaba Group's product technology capabilities and operation data capabilities, forming a strong support for each front office business. Juhuasuan and 1688, which were subsequently launched, all benefited from the construction of middle office.

Alibaba Group's system architecture of "big middle office, small front office" has contributed to the reform of enterprise business architecture. It is not only the result of Alibaba Group's development experience of Internet business, but also the summary of advanced experience in industry development. Therefore, it is also worthy of being exported to industry users. As Tencent, JD, Meituan and other Internet head enterprises began to promote the organizational structure reform from the group level to build middle office as the goal, a number of startup companies in data field also proposed the middle-office strategy, which made more and more enterprises pay attention to the middle office. At present, the development of middle office is still at the primary stage of application, in which the head enterprises in the industry are utilizing and other enterprises are trying or waiting. Under the background of the continuous advancement of China's digital process, the digital transformation of enterprises is entering a new stage, and the role of the middle-office system is becoming more and more important. It mainly includes the following aspects:

(1) The industry is at the stage of research and practice, and it still takes time for wide application

The middle-office system is now at its early stage, and the percentage of enterprises that actually build or upgrade the middle-office system is very small. There are no particularly mature methods and tools in the consumer-oriented industry, and different manufacturers are still in exploring and practicing phase. The commercial value of the middle-office system has not been fully demonstrated. The wide application still needs to wait for the development of the market, and there are still many challenges in the implementation.

(2)   Enterprises have a high degree of attention, but lack awareness of the whole or relevant concepts

Extensive consumption area and industrial Internet enterprises pay more attention to the development and related dynamics of the middle-office system. After Alibaba, Tencent, JD and other enterprises announced the construction of the middle-office system, most Internet enterprises began to pay attention to it, especially those enterprises that need to quickly respond to the change of market demand.

Although, in the past two years, the Internet heads and some of the industry's head enterprises in China have begun the construction of middle-office system, the whole market, especially the traditional industry, has a relatively shallow cognition to it. Only a few enterprises have risen it to the strategic level, most enterprises remain doubtful about the need and the practical usefulness of middle-office system, the market also needs further popularization.

(3)   The construction cost of the middle-office system is high, the construction cycle is long, and it needs to be built step by step

Benefiting from the development and wide application of new technology in recent years, enterprises have a rational understanding of the cost of technology practice. The construction of the middle-office system cannot be accomplished overnight, and its integration with the original mechanism of the enterprise will be a long-term process. Its construction cost is over one million yuan, and its construction cycle is calculated in months or years.

The whole construction process needs to be accomplished step by step according to the business objectives of the enterprise, so as to realize the gradual evolution of the digital capabilities of the enterprise and the continuous superposition of value. Meanwhile, in the process of construction, it is also necessary to train the operation and maintenance management team for customers, and even reconstruct the entire IT team, so as to improve the digital operation capability of the enterprise.

(4)   At the beginning, the competition pattern of the market is not clear, and the head enterprises in the industry become the focus of competition

The service providers of middle-office system are mainly composed of three kinds of manufacturers: Internet head enterprises, independent developers of middle-office system and integrated service providers of digital enterprise. The whole industry of middle-office system is still at the early stage, the market concentration is low, the pattern has not yet formed, each factory is in the horizontal and vertical expansion through their own channels.

New players continue to enter the market, and with the help of capital, competition in some segments of the industry will become increasingly fierce. The more data-sensitive industries such as Internet, finance and new retail will become the key areas for service providers to compete, and the head enterprises in the industry will become the focus of competition.

Due to the business volume and data complexity of the head enterprise itself, the requirements for supporting data security management are high, so the requirements to service providers are also very high.

Therefore, it is very important for service providers to get orders from the head customers, which is the side endorsement of their comprehensive strength and has an immediate effect.

(5)   It is extending to different fields to further facilitate the digital transformation of enterprises

The middle-office system is extending from the Internet industry and new retail industry to other industries and subdivided application fields, exploring specific solutions while further mining the industry attributes.

Especially for service providers focusing on a specific subdivision field, they are exploring the middle-office system of such subdivision industry, because in the vertical scenarios, this kind of service providers can better serve the customers of specific type, and provide a more targeted industrial solution, which can better help the customer of such subdivision industry in digital transformation and upgrading.

In addition, in the process of extension, cooperation between different companies can form a closed loop of service ecology solutions in a particular field.

4.2.2   Main Applications of Middle-Office Technology

1.   Extensive retail

At present, it is the field with the largest number of middle-office systems and the fastest development. Driven by technology and consumption upgrading, digital transformation has become the consensus of the whole extensive retail industry. Especially in recent years, due to the mobility of "people", customization of "goods" and diversification of "fields", "people" as the center, enterprises have promoted full-chain business and experience optimization from consumers to retailers and brand owners with finer granularity, realizing "data business integration". Understanding data is understanding the demand of consumers, thus relying on middle-office system, the enterprise will capitalize the original entity data and conduct thorough data flow, so as to promote the overall data capability of the retail enterprises, quickly understanding consumer demand and quickly responding, realizing the integration of the online and offline business, ultimately improving consumer experience and forming a new retail ecosystem. Extensive retail enterprises are closer to consumers, so the construction of the middle-office system is often more vertical, focusing on the scenario and personalized customization, and striving to achieve closed data loop of consumer-centered online and offline integration, accurate analysis of consumer demand, in order to respond to the increasingly rapid changes in the market terminal.

2.   Finance

To better cope with the changes, middle-office technology has enabled financial institutions to transform and upgrade. In the traditional financial system, IT system is usually constructed by each business department according to the business line. At present, there are hundreds or even thousands of systems in each organization, forming the current crisscross IT system matrix. With the transformation of front-end finance from sales type to service type, there are more and more business scenarios with high concurrency, large amount of data, strong consistency and horizontal expansion capability. More and more organizations need to provide more personalized products under the premise of security and control, in order to find a feasible mode for differentiated operation. Meanwhile, under the situation of strengthening supervision and unified risk control, the service capability and operation capability of IT facilities are increasingly required. In the digital transformation of financial institutions, the middle-office system has become the main plan to realize the omni-channel and all-chain agile business capability. Financial institutions have a large number of users and data to expand their business in the upstream, middle and downstream sectors of various industries. Only in a real financial scenario can changes in policies, rules, and requirements be addressed more quickly through internal departmental connection and end-customer connection. Financial institutions have shifted from business-centered to people-centered, creating financial service mode of scenarios, and even stimulating new demands and creating new business modes. At present, compared with the business middle office, the data middle office is relatively more applied in the financial field.

3.   Digital government

The middle-office architecture is a new path for government informatization, promoting the development of Digital Government 2.0. Digital government is the final manifestation of the digital transformation of government services under the background of digital China. It is the driving power of promoting government reform and social innovation and development, as well as the construction of digital China. Over the past years, government informatization has been continuously promoted and deepened, the construction of Digital Government 1.0 has been initially completed, and the online, networked and mobile government services have been realized. During this period, the government has built a complete information framework for government services, initially formed a new technical architecture system of government cloud infrastructure, and improved the office efficiency of government clerks. From Digital Government 1.0 to Digital Government 2.0, it focuses on solving the problem that the innovation in government service business is far to meet the social demand, ensuring the interconnection of data between various departments, promoting the integration of data and business, ensuring the value of "data" to enable "business" services, and improving the supply-side capability of service-oriented government. The application of the middle-office system has realized the data-oriented operation of government services and the process reengineering of government departments, comprehensively improved the government's capability to provide convenient services, excellent social governance and scientific decision-making, and brought a new governance and service mode to the government. The reform of online services to public has been finally realized.

4. Medical care

In the medical field, middle office is still at the concept stage, and it will drive the integration of regional medical data in the future. The informatization of medical institutions began in 1999. After 20 years of development, hospitals have built multiple information systems around business modules, such as inpatient, outpatient, nursing and testing. However, the manufacturers of these systems are different, resulting in the isolated business data in the business system and lack of compatibility and integration. At a macro level, data of each hospital system cannot be transferred between hospitals or regions. With the comprehensive transformation of the medical and health service system, the existing HIS, EMR and other systems cannot well reflect the value of medical data, and the medical information system architecture is faced with upgrading to adapt to the complex and changing needs. In the future, driven by medical treatment and management, hospital informatization based on the middle-office system will break through medical information walls, precipitate effective medical data, combine refined operation, clinical path and single disease management, and connect the service chain, business chain and data chain to realize the data analysis of various hospital subjects such as clinic, scientific research and operation, so as to facilitate doctors' operation implementation, improve the diagnosis and treatment effect, and enhance the working efficiency of the hospital. Through access to inter-hospital data, the utilization rate of medical resources can be improved to achieve efficient and high-quality medical services, and it will be helpful to the implementation of policies such as medical alliance and tiered diagnosis and treatment as well as reconstruction the value system of hospitals.

4.3 Applications of Digital Middle Office in Smart Port

4.3.1 Significance of the Concept of Middle Office to the Integration of Port Resources

Port enterprises and e-commerce enterprises have completely different business modes and operation models, but they face many similar or even the same problems in the process of digital transformation. Taking a port enterprise in China as an example, it had a major consolidation of the logistics sector in 2009. Almost all the rear yards, yards for container scheduling, ro-ro auto yards, ore yards and a large number of waterless ports and inland yards outside the port were integrated and reorganized into a secondary company, which specialized in the logistics sector of the port. Years of practice has proved that this is undoubtedly a very successful and necessary integration from the perspective of the overall business strategy, but during the integration, the enterprise still inevitably encountered some problems in information construction. In the process of information integration, due to the limited manpower, financial resources, only the intensive integration of the storage yard system within the port area was considered. However, for different yard operating branches, especially the waterless port branch, relatively independent plans based on the business needs of their own departments were put forward. The IT department built complex and partially repetitive stovepipe systems to meet different (and sometimes conflicting) business needs. The construction of stovepipe systems not only brought about the repetitive construction of functions, but also the repetitive maintenance, which led to the repetitive investment of enterprises. In order to break through the stovepipe systems, it is necessary to design a third-party integration plan or introduce the concept of enterprise service bus (ESB). The integration cost and collaboration cost are high, so the digital integration of the yard system has not been completely finished. During the construction and introduction of new system, although each department builds customized optimal solutions according to its own business needs, these solutions may be only partial optimal. If the company is seen as a whole, it may not be the globally best solution. Therefore, if the system is not built from a global perspective, and the existing system is not reformed, upgraded and reused, then it can only introduce new complexity on the old complexity again, resulting in increasingly more complex but less efficient system.

4.3.2 Development Foundation of Port Middle-Office System

The port business is complicated, with crisscrossed data in huge volume. The port enterprises keep tracking and actively practice the emerging technologies in the fields of informatization, digitalization and intelligence. Large ports in China are all committed to port digital transformation, from cloud computing and data warehouse to big data center, artificial intelligence and blockchain.

1. Construction foundation of cloud computing platform in port

In recent years, the supporting ability of the port information infrastructure has been significantly improved. Large port operators in China, such as Tianjin Port Group, Shanghai Port Group, Ningbo Port Group, Shandong Port Group, have built cloud computing centers of different scales to varying degrees, constantly promoting the core on cloud, and gradually realizing the intensive management of internal operation system. Obviously, the construction of port cloud centers based on cloud computing has become a trend. For instance, a large port in China conducted a three-year program to "improve the weakness" from 2017 to 2019, with the goal of digital transformation. Through unified core TOS, the cloud computing center of group level has been constructed. It has become the first port in China to deploy TOS and other core operation systems in an intensive manner. At present, the construction magnitude and usage ratio of cloud computing centers have taken the lead among the ports in China. The cloud computing capacity has reached 7116 Vcpu, the structured data storage capacity has reached 1.86 PB, and the average daily growth of structured data has reached 14 GB. The proportion of the core system on cloud continues to increase, and the aggregation effect of big data is significantly enhanced. In particular, the service capability of cloud center system and the common sense of intensive management are more advanced. Companies have gradually become more inclined to management and control of "cloud mode".

2.   Construction foundation of big data in port

All ports are actively exploring the integration and application of big data. The level of data resource integration is mainly reflected in the volume and integrity of resource collection of big data center, and in the integration and use of resources based on port digital middle office. The data interaction between the port and the customs, maritime administration, municipal transportation commission and other regulatory departments has been continuously deepened, and the port EDI system upgrading, global shipborne AIS data integration, information integration of national road transportation vehicles and other data integration have been gradually completed. With government support, the ports have participated in the construction of major government projects such as the simplification of customs clearance procedures, network supervision system for the special region, and the electronic payment system of ports, a local electronic port platform connecting port supervision departments including customs, inspection and quarantine, relevant departments of local government including foreign trade, industrial and commercial administration, as well as major financial institutions at home and abroad has been built, with service network covering seaports, airports and all specially supervised areas, and radiating to surrounding areas. Large port enterprises have relatively complete channels to acquire port and shipping data, with large data volume and scale, and are gradually exploring deep integration and effective utilization of big data. Ports in China are trying to build up a data resource integration platform, and the government supervision departments are actively cooperating with the construction of smart port, to help build the port and shipping data resource integration platform with more complete data, more perfect function, more efficient operation and more intelligent control. The degree of data interaction between port and customs and the innovative interaction mode of embedded control are constantly developing, which has played a positive demonstration role in the industry. In general, the integration and application of big data in all ports are at the stage of data resource integration, and the standardized and large-scale application of big data based on advanced concepts such as digital middle office is still being explored.

To sum up, the rapid development of these technologies and their integration with real business have catalyzed the middle-office strategy. The concept of port middle office has sprouted, and some large ports have begun to research, explore and plan port middle-office system suitable for their own needs.

The port middle-office system is the transformation of IT architecture of port enterprises. From the global perspective of the port group, the port middle-office system is planned from the aspects of overall strategy, business support, customer connection and operation innovation. Therefore, the port middle-office system includes the main business of the port and is more focused on how to support the online business. In fact, in the process of exploring the "online + offline" strategic mode, some port enterprises in China have more or less consciously generalized and created some common service, that is the basic concept of business middle office. At the beginning, the port enterprises were led by their container terminal companies to explore and build online business reception platforms that were convenient for customers to use, realizing online processing of container delivery, payment, container collection and other businesses. Some general cargo and dry bulk cargo terminals have also made attempts. These platforms were independent stovepipe systems. However, in essence they were all online business services such as container business reception and online payment. Since they were online trading platforms, they all involved the processes of information verification, business reception and payment, which had common needs. Therefore, in recent years, some large ports in China have gradually developed unified public service platforms, such as online business reception center within the whole port, which could connect with the operation system of the terminal company subordinate to the port group, and have unified interface service standards. However, there is still a lot of room for improvement at the level of generalization, reusability, and open degree for these platforms. Hence, the port business middle office is in the embryonic state in the port industry in China and even in the world.

4.3.3 Construction of Port Middle-Office System

Taking the port online business reception center as an example, the core of the middle office is online business reception. The object is the container/cargo related services, which are sold to customers through terminals, storage yards and other operating enterprises, and the transaction voucher is the reception plan. Online transactions require payment; pre-sale marketing activities are required to attract customers for online business processing; customers will evaluate the terminals and yards that provide handling/transportation services. It can be seen that a typical business middle office consists of multiple business and service centers, including customer management center, container/cargo information center, online transaction center, complaint and evaluation center, terminal/yard enterprise center, payment center, marketing center and logistics inquiry center, etc..

1.   Customer management center

The customer management center serves the whole life cycle of customer business and provides customers with specific rights and services. The customer retaining team can interact with customers through the enterprise membership center to improve the service quality. The main capabilities include the followings.

① Customer retaining and management, including user registration, customer company registration, company operator management, personal information maintenance, user cancellation and other related capabilities.
② Customer system management, including the establishment of customer system, rules of credits, rules of rewards and punishments, grades, rights and interests and other relevant capabilities.
③ Customer service management, including customer addition, import, query and other related capabilities.
④ Credit management, including credit acquisition, verification, reset, freezing, exchange and other related capabilities.

2. Container/cargo information center

The container/cargo information center provides the management capability of containers and cargoes that can be processed online, and constructs business related data around online business, such as container no., bill of lading no., cargo information, business type, service mode, service location, basic rate, etc. The main capabilities include the followings.

① Cargo property management, including the maintenance and query of the properties such as container no., bill of lading no. and inspection application no., as well as the management of properties and property groups.
② Business data management, including the capability to create, edit, query and disable data such as business types, business service modes, business standard rates, etc.

3. Online transaction center

The online transaction center is responsible for the overall life cycle management of business transaction orders, including service selection → order generation → consolidation and split → circulation → payment → handling/transportation operations → complaint and other issues handling → completion. The core system of all online businesses is built around transaction orders. The main capabilities include the followings.

① Service selection, including service addition, value-added service selection, editing, query, verification and other related capabilities.
② Forward transaction management, including transaction order generation, initiating payment of transaction order, pushing logistics node status of container/cargo, door-to-door self-delivery, verification and cancellation and other related capabilities.
③ Backward transaction management, including order cancellation, reservation cancellation and other related capabilities.
④ Order data management, including transaction orders, payment records, operation records, refund records and other data management capabilities.
⑤ Transaction process choreography, including user-definition and configuration of transaction process nodes, making it easy to deploy processes according to business requirements.

4.   Complaint and evaluation center

The complaint and evaluation center provides the capability to complain and evaluate the main object of the service business, as well as the management capability of evaluating operation, so as to meet the needs of evaluation users with different roles to release and add evaluation as well as for platform review and appeal. The main capabilities include the followings.

① Complaint management, including management of the complaint object, complaint rules configuration, complaint level and other related capabilities.
② Evaluation management, including management of evaluation object, evaluation rules configuration, evaluation level and other related capabilities.
③ Customer online evaluation capability, including the release, modification, addition, reply and other related capability of evaluation.
④ Evaluation supervision capability, including released evaluation review, appeal review, shielding evaluation and other supervision capabilities.

5.   Terminal/yard enterprise center

The terminal/yard enterprise center provides online enterprise main management, type management, business object management and other capabilities to support the group to provide online stores for enterprises, as well as enterprise management, enterprise membership, membership level management, etc. The main capabilities include the followings.

① Enterprise information maintenance and management, including the capability of opening online enterprise, review and maintenance of the basic information of the enterprise.
② Enterprise operation management, including online staff management, right management and other related capabilities.

6.   Payment center

The payment center provides standard payment services for service enterprises such as terminal/yard, including payment and collection, financial reconciliation and other services. It can stably output WeChat, Alipay, UnionPay and other payment functions by connecting with multiple mainstream channels. The main capabilities include the followings.

① Payment capability, including the creation of payment order, notification of receiving channel, query channel order and other basic payment capabilities.
② Payment routing, including payment channel management, payment method management, payment merchants and application opening management and other related capabilities.
③ Capital account, including capital account management, recharge maintenance, withdrawal and other related capabilities.

7.   Marketing center

The marketing center provides full chain management of online business-related activity planning, declaration, approval, execution and verification, as well as basic promotion capabilities, such as value-added promotional activities and discount offers. The main capabilities include the followings.

① Activity template management, including providing strategy template of marketing activities, rule configuration, conditions, action template and other related capabilities.

② Activity management, including providing basic information configuration of specific activities, triggering conditions and other related capabilities

③ Discount management, including providing dynamic query of discount conditions, enabling and disabling discounts and other related capabilities for customers with large business volume and customers with high participation in online processing.

8.   Logistics inquiry center

The logistics inquiry center provides container/cargo port operation condition, logistics and transportation condition inquiry and other related service capabilities. It mainly includes the followings.

①   Port operation condition management, including condition of service enterprise, container blocks, bays and related management capabilities.

②   Logistics condition management, including logistics demand, cost, logistics condition monitoring and other related capabilities.

To sum up, the construction of port middle-office system can be used in the development, design and service of multiple online platforms simultaneously. Therefore, middle office can save system construction and operation costs for port enterprises that simultaneously build and operate multiple online business platforms. The middle office can not only avoid the repeat of function construction, but also increase the flow and promote each other through the all-channel access to customer systems, and also reduce the operation cost and personnel. With the gradual maturity of middle-office technology and the continuous refinement of the requirements of port middle-office system, the middle-office system will become an important part of the construction of smart ports.

Bibliography

1. Thönes J (2015) Microservices. IEEE Softw 32(1):116–116
2. Zhong H (2017) The way to transform enterprise IT architecture. China Machine Press

Chapter 5
Smart Port and Blockchain Technology

Since the concept of blockchain was put forward in 2008, its application has gradually evolved from cryptocurrency to platform providing trusted services. All walks of life are actively exploring the innovative industrial application mode of "blockchain plus".

Effective organization and coordinated logistics chains across industries, departments and regions are important symbols and external embodiment of a smart port. Smart port will focus on the port value service chain, actively explore business reform and service innovation, and enhance the comprehensive soft power of the port. The technical features of blockchain, such as decentralization, process verifiability, traceability, tamper-proof and openness, can help to innovate the port development pattern and build an open, collaborative and highly interconnected port ecosystem system.

5.1 General Introduction to Blockchain

As an important part of the information technology system, blockchain is expected to become another technology that changes the development mode of human society and economy after steam engine, electricity, information and Internet technology. Blockchain is a new technology, which integrates many existing interdisciplinary disciplines, including mathematics, cryptography, computer science, etc. Due to the interdisciplinary integration support, blockchain has built a self-governing, reliable and traceable system in the digital world.

5.1.1 Blockchain Concept

Blockchain originated on November 1, 2008, a person or group named Satoshi Nakamoto published a paper titled *Bitcoin: A Peer-to-Peer Electronic Cash System*

© Shanghai Scientific and Technical Publishers and Springer Nature Singapore Pte Ltd. 2022
W. Mi and Y. Liu, *Smart Ports*, https://doi.org/10.1007/978-981-16-9889-7_5

in the cryptography mailing list of metzdowd.com. This paper pointed out that the blockchain technology was the technical foundation to build the bitcoin system, which could record all metadata and encrypted transaction information, thus creating an electronic cash system implemented entirely through peer-to-peer (P2P) technology. The system would allow online payment parties to trade directly without requiring a third-party intermediary.

From the perspective of protocol, blockchain is a kind of Internet protocol to solve the problem of data trust. From the economic perspective, blockchain is an Internet of value that promotes the efficiency of cooperation. From the accounting perspective, blockchain is a distributed accounting technology or accounting system, which is a ledger in the form of electronic records. Each block is a page of the ledger and "linked" from the first page to the latest page. Once these blocks are confirmed, they are almost impossible to modify, and each block contains all transactions in the current period of time.

In the *White Paper on the Development of Blockchain Technology and Application in China (2016)* released by the Ministry of Industry and Information Technology of China, blockchain was defined as a new application mode of computer technologies, including distributed data storage, point-to-point transmission, consensus mechanism, encryption algorithm, etc. It is a decentralized and trustless infrastructure and distributed computing paradigm.

5.1.2 Blockchain Types

According to the degree of openness, blockchain can be divided into three types: public blockchain, federated or consortium blockchain and private blockchain.

A public blockchain means that any individual or group sharing a blockchain can send a transaction on it as long as it is connected to the chain, and the transaction can be effectively confirmed by the blockchain, and any group or individual can participate in the consensus process. The public blockchain is the first to appear, and it is also the most widely used at present. It is mostly used in the scenarios of deregulated, anonymous and free cryptocurrencies such as Bitcoin. Such blockchains are considered "completely decentralized."

A federated or consortium blockchain is a blockchain in which the consensus process is controlled by certain pre-selected nodes. A number of pre-selected nodes are first designated with the right to validate the block within the industry collective, and the generation of each block is jointly decided by all the pre-selected nodes. Other nodes can only access the blockchain to be responsible for transactions, but do not participate in the consensus process. Anyone can conduct limited queries through the open application interface of the blockchain. Such blockchains are considered "partially decentralized."

Private blockchain is to only use blockchain technology for accounting operations, and not open to public. The object can be either a company or an individual that has written access to the blockchain alone, perhaps with highly restricted access to the public.

5.1.3   Supporting Technologies of Blockchain

Blockchain is based on consensus mechanism, distributed data storage, P2P transmission, encryption algorithms and other information technologies, driving the transformation of the Internet from an information Internet to a value Internet.

1.   Consensus mechanism

The so-called consensus refers to the process in which multi-participant nodes reach agreement on some data, behaviors or processes through the interaction of multiple nodes under preset rules. Consensus mechanisms are the algorithms (consensus algorithms), protocols, and rules that define the consensus process.

Consensus algorithms include Proof-of-Work (PoW), Proof-of-Stake (PoS), etc. In PoW mechanism (commonly known as mining, each node called a miner), each node contributes its own computing resources in competition to solve a mathematical problem with dynamically adjusted difficulty. The miner who finds a valid hash and successfully solves the math problem gets the accounting rights to the block, and all transactions in the current period are packaged into a new block and linked to the main chain in chronological order. When a miner finds this value, it shows that the miner has indeed done a lot of calculation, that is, the PoW is obtained. Meanwhile, the PoW mechanism has significant defects, and the resource waste (such as electricity) caused by its computing power has always been criticized by researchers. PoS mechanism is an alternative scheme to solve the resource waste and security defects of PoW mechanism. In essence, PoS consensus uses proof of stake to replace proof of work based on hash computing power in PoW, and the node with the highest number of staked coins in the system rather than the highest computing power obtains the right to validate the block. Stake is represented by the ownership of a node in a specific number of coins, known as coin days. The coin days are the product of a given number of coins and the length of time they are last traded. Each transaction will consume a certain number of coin days. The transaction for each block is submitted to the block for the coin days it has spent, and the block with the highest cumulative coin days spent will be linked to the main chain.

2.   Smart contracts

The idea of smart contract first emerged in 1994, around the same time as the Internet. The term was coined by cryptographer Nick Szabo with the definition that "a smart contract is a computerized transaction protocol that facilitates, validates, or executes the negotiation or performance of a contract, or makes the terms of a contract unnecessary".

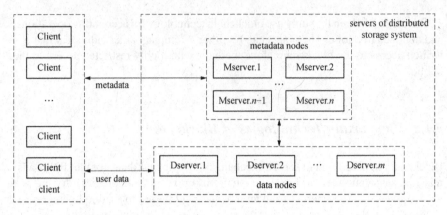

Fig. 5.1 General architecture of distributed storage system

The emergence of blockchain technology makes smart contract active again, and is considered to be another hot technology applied in blockchain technology. Smart contract is redefined as event-driven, stateful program running on a ledger that can be copied and shared, and capable of holding assets on the ledger.

3. Distributed storage

Distributed storage means that the data are distributed to several independent servers in the network, and the storage software is run for management, so that the servers can be integrated as a whole to provide storage services. In this architecture, functional tasks of the original system are decoupled into metadata service (including data attribute maintenance, storage location addressing, rights management, node management, etc.) and data service (including reading, writing, deleting, altering of data, etc.), respectively borne by the metadata nodes (Mserver) and data nodes (Dserver). The general architecture is shown in Fig. 5.1.

4. P2P network technology

P2P network technology, also known as peer-to-peer technology, is an Internet system without a center server and relies on user groups to exchange information. Since there is no center server in P2P network, it has the advantages of attack resistance and high fault tolerance. In addition, each node has the equal status, and the service is scattered on each node. Therefore, the attack to some nodes or network has almost no impact on the whole system.

5. Hashing algorithms

Hashing algorithms map input data of arbitrary size to a hash of a fixed size. For example, the SHA256 algorithm maps input data of arbitrary size to fixed size output of 256 bits, and this binary value is called a hash. A hash function algorithm is designed to be a one-way function, infeasible to invert. When applying the hash function, it takes about the same amount of time for inputs of different sizes and produces fixed-sized outputs, and even a single bit difference in the input can result in significantly different output values. The hash values of the data verify the integrity of the data and hash algorithms are commonly used for quick lookup and in encryption algorithms.

Hashing algorithms are widely used in blockchain. Blockchains usually store the hash values of the data instead of the original data. Node information in Merkle tree is obtained by running twice SHA256 hash algorithm.

6. Merkle tree

Merkle tree is a tree with hash-based architecture invented by Ralph Merkle. Its data structure is a tree, usually binary, or multi-branched. The leaf node is the hash value of the data block, and a non-leaf node is a hash of its respective child nodes. Merkle tree is an important data structure of blockchain, which is used to quickly summarize and verify the existence and integrity of block data.

7. Asymmetric encryption

Asymmetric encryption is the encryption technology integrated into the blockchain to meet the security requirements and ownership verification requirements. Asymmetric encryption requires a pair of keys, namely, public and private keys. The public key is open and the private key is confidential. The information encrypted by the private key can only be decrypted by the corresponding public key, and the information encrypted by the public key can only be decrypted by the corresponding private key, that is, the public key is used for encryption and the private key for decryption; the private key is used for signature, and the public key for verification.

5.1.4 The Block of a Blockchain

The words "block" and "chain" in a blockchain are both terms used to describe the characteristics of its data structure. In a blockchain system, transactions are organized into blocks, which are then organized into logical chains. From a data perspective, a blockchain is a chain-like data structure that connects blocks of data in chronological order. A block is a structural data unit that consists of two parts, the block header and the block body, as shown in Fig. 5.2. The block header holds all kinds of information to connect to the previous block, information to validate the block, and timestamps, etc. "Version" is used to track software/protocol upgrades, "previous block hash" is a reference to the hash of the previous (parent) block in the chain, "timestamp" is the approximate creation time of this block, "nonce" is a counter used for the PoW algorithm, "difficulty target" is the PoW difficulty target for this block, "Merkle root" is a hash of the root of the Merkle tree of this block's transactions. The block body contains all the transaction information in the block. The number of transactions is used to declare the specific number of transactions in the block body. The transactions are organized into a Merkle tree structure, where the

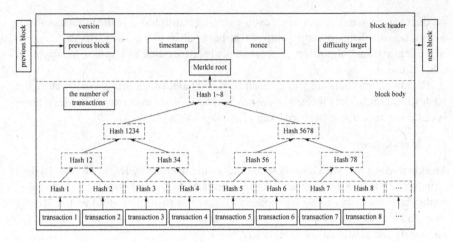

Fig. 5.2 Block structure of the blockchain

transactions are stored on the leaves of the Merkle tree, and the hash is pound-merged until the root node is obtained.

Genesis block is the first block in the blockchain system. From the genesis block, all blocks are connected in series to form a blockchain. Figure 5.3 is a chain connected by three Bitcoin blocks.

5.1.5 Workflow of Blockchain

The main workflow of blockchain is divided into 5 steps: new transaction creation, P2P network communication, PoW, network verification, transaction writing into the ledger.

(1) Nodes create a new transaction and broadcast the new transaction to the entire network;

(2) The receiving node verifies the received transaction. If it is valid, it will incorporate the transaction into a new block;

(3) All miner nodes in the whole network (nodes in the network that have the ability to package and verify transactions) perform consensus algorithm on the above block and select the packaging node;

(4) The packaging node uses consensus algorithm to broadcast the new block it packs to the whole network;

(5) Other nodes verify the block of the packaging node and, after several verifications, add the block to the blockchain.

block height:	419197						
hash value of header:							
00000000	00000000	01d1c177	c99320ca	2bda79b8	a90f71c1	d4fde2ec	0ad884a7
Previous	Block						
00000000	00000000	00d803b1	0293a498	0950a29c	0f48d86c	ffb669a3	509206b5
Timestamp		2016-07-04		02:57:45			
Difficulty		209,453,158,595.38					
Nonce		686437237					
Bits		402997206					
Merkle Root							
1ee603f6	0573d600	2c9af11b	c0360b07	9ed646b4	2abbd839	1b4d9c25	472b4351

⇧

transaction

block height:	419198						
hash value of header:							
00000000	00000000	04df28b7	d15e3512	0d3e50e2	ae0a245c	d322ac97	dd0d2831
Previous	Block						
00000000	00000000	01d1c177	c99320ca	2bda79b8	a90f71c1	d4fde2ec	0ad884a7
Timestamp		2016-07-04		03:14:01			
Difficulty		209,453,158,595.38					
Nonce		3.59E+09					
Bits		402997206					
Merkle Root							
2adba499	1559df32	bea5161e	0e4f051c	949b5fee	9388e958	31684bd4	c1a0dddf

⇧

transaction

block height:	419199						
hash value of header:							
00000000	00000000	017ba0d3	cd49cd63	a9c36c1c	066899a6	77c8c87b	cc550914
Previous	Block						
00000000	00000000	04df28b7	d15e3512	0d3e50e2	ae0a245c	d322ac97	dd0d2831
Timestamp		2016-07-04		03:15:05			
Difficulty		209,453,158,595.38					
Nonce		1672094155					
Bits		402997206					
Merkle Root							
8fad5acd	9326d0e4	09d23857	e2dd25fd	89fdec64	97d082c3	fe51eb0c	2f585c09

⇧

transaction

Fig. 5.3 A chain connected by three Bitcoin blocks

5.1.6　Features of Blockchain

The features of blockchain, such as decentralization, data transparency, transaction traceability, collective maintenance, security and credibility, and contract autonomy, ensure that transaction activities can be carried out anywhere and anytime, breaking through the limitations of traditional trade in space and time, and creating more trading opportunities for both parties.

There is no center hardware or management organization in the whole system. The rights and obligations of any node are equal, and the damage or withdrawal of any node will not affect the operation of the whole system.

The technical foundation of blockchain is open source. In addition to the private information of transaction parties being encrypted, the data of blockchain are open to everyone. Anyone can query the blockchain data and develop related applications through the open interface, and the operating rules of the data and system are open and transparent blockchain uses timestamp, consensus mechanism and other technical means to make the data tamper-proof and traceable. It applies a chain-type block structure with timestamps to store data, thus adding the time dimension for data, which have strong verifiability and traceability. It provides technical support for the establishment of cross-organization traceable system.

Any node in the blockchain can participate in the system maintenance. While participating in the recording, each node also verifies the correctness of the recording results of other nodes, and each node can obtain a copy of the complete database (blockchain).

Blockchain technology adopts the principle of asymmetric cryptography to encrypt data. Meanwhile, it relies on the computing power formed by consensus algorithm to resist external attacks and ensure that blockchain data cannot be tampered or forged, with high security.

The predefined business logic enables nodes to be autonomous based on highly trusted ledger data, automating the execution of services between person-to-person, person-to-machine, and machine-to-machine interactions.

5.2　Development of Blockchain

5.2.1　Evolution Path of Blockchain

The American scholar Melanie Swan, in his book *Blockchain: Blueprint for a New Economy*, divided the history of blockchain into three important stages according to its significant influence on various application fields, namely, Blockchain 1.0 (programmable currency, 2009 to 2014), Blockchain 2.0 (programmable finance, 2014–2017) and Blockchain 3.0 (programmable society, 2017 to present).

(1) The stage of Blockchain 1.0 is mainly the era of digital currency, represented by the application of Bitcoin, which was the application of cryptocurrency. The

construction of decentralized digital payment system has realized the diversified financial services such as fast currency transaction and cross-border payment.

(2) The stage of Blockchain 2.0 is mainly the application of smart contracts, typical representatives are Ethereum and Hyperledger. The application of blockchain has extended to the financial field, which was the decentralization of the smart asset and smart contract market, and could be used for the transfer of digital assets beyond currency.

Ethereum is the first and currently the most active public chain of Blockchain 2.0 in the world. Ethereum is a programming platform that provides a variety of templates that users can simply link together to build their own applications. As a result, building applications on Ethereum has become much cheaper and faster, which has made Ethereum one of the best projects in the blockchain.

Hyperledger is an open-source project launched by the Linux Foundation in 2015 to advance blockchain technology. It leads global industries, including finance, the IOT, supply chains, manufacturing and technology, to collaborate on blockchain technology and build an open platform to meet different needs from various industries and streamline business processes.

(3) The stage of Blockchain 3.0 is mainly an era of comprehensive application of blockchain. Blockchain technology allocates global resources in a decentralized way. At present, healthcare, IP copyright, education, culture and entertainment, communications, philanthropy, social management, sharing economy, the IOT and other fields are implementing blockchain application projects, "blockchain plus" is becoming a reality.

5.2.2 An Overview of Blockchain Development

As the rise of digital currency represented by Bitcoin, the focus of several government departments, financial institutions and Internet giant companies on its underlying blockchain technology continued to heat up, many countries have recognized the blockchain technology has huge application prospect, and started thinking about the development path of blockchain from the country level. China also has started the research and practice.

1. Development in the world

In August 2013, Germany announced the legal status of Bitcoin, and it had been included in the national regulatory system. The German Banking Association believed that blockchain technology could have a major impact on financial markets. In 2016, BaFin explored the potential applications of distributed ledgers, including their use in cross-border payments, transfers between banks and the storage of transaction data.

In December 2013, the world's first Bitcoin ATM was put into use in Vancouver. Canadian dollars deposited in the ATM by the users could be converted into Bitcoin and then transferred to a Bitcoin account on the network. Users could also exchange

Canadian dollars directly from their own Bitcoin account. In June 2016, the Bank of Canada demonstrated an electronic version of the Canadian dollar, CAD-Coin, developed by using blockchain technology.

On January 26, 2015, Coinbase, which was invested by the New York Stock Exchange, was approved to establish a Bitcoin exchange, and the legislative process of Bitcoin regulation was initially completed in the United States, represented by the State of New York. In June 2015, the New York Department of Financial Services released the final version of BitLicense, the regulatory framework for digital currency companies. The US Department of Justice, the American Stock Exchange, the US Commodity Futures Trading Commission, the US Department of Homeland Security and other regulatory agencies showed their support for the development of blockchain technology from their respective regulatory fields.

In January 2016, the UK government released a research report on blockchain titled *Distributed Ledger Technology: beyond block chain*, suggesting that blockchain should be included in the national strategy of the UK and promoted in finance, energy and other fields. In June 2016, the UK government launched a blockchain pilot to track the allocation and use of welfare funds.

2.   Development in China

In December 2016, blockchain was listed in the national informatization plan in the *13th Five-Year Plan for National Informatization*, and identified as a strategic frontier technology, which marked the beginning of China's efforts to promote the development of blockchain technology and applications. According to incomplete statistics of *Knowledge Economy*, by December 2019, more than 20 policies had been issued at the national level to promote the standard establishment and application of blockchain. The policy makers included the State Council, the Ministry of Industry and Information Technology, the Ministry of Commerce, the Post Bureau, the Central Bank, the Ministry of Education, and the Cyberspace Administration of China, etc.

According to the "Global Blockchain Enterprise Invention Patent Rankings 2019 (TOP100)", by October 25, 2019, seven of the top ten applications for blockchain technology invention patents were from China, among which Alibaba ranked first with 1,005 applications. In the two batches of blockchain information service record lists released by the Cyberspace Administration of China (CAC), Alibaba has registered three products, namely Blockchain as a Service (BaaS) platform under Ant Financial, Ali Cloud Blockchain Service under Ali Cloud and Hundsun Sharing Ledger (HSL) of Hundsun Technologies.

5.3   Typical Applications of Blockchain

Since January 2009, Nakamoto developed the first client program to realize the Bitcoin algorithm, namely the official announcement of the birth of Bitcoin, the applications of blockchain have been extended from a single digital currency application to all fields of economic society, mainly including financial services, supply chain management, culture and entertainment, education and employment, etc. At present, the more mature application scenarios are financial services, and other industries have also carried out exploration and practice in the applications of blockchain technology.

5.3.1 Blockchain + Financial Services

1. Digital currency

Blockchain technology, as the support behind digital currency, has attracted the attention of financial giants, including Citibank, J. P. Morgan Chase, Goldman Sachs, Bank of New York Mellon, HSBC, Barclays Bank, etc. Many financial giants have cooperated with blockchain companies to study the applications of blockchain technology in the financial sector. In 2016, J. P. Morgan Chase, Barclays Bank and other dozens of world-renowned banks joined the R3 Blockchain Alliance to explore and study the application scenarios of blockchain in the banking industry. In February 2019, J. P. Morgan Chase Bank released MorganCoin, a cryptocurrency used for inter-agency clearing. In March 2019, IBM announced that the blockchain of cross-border payment, World Wire, was officially put into actual operation. On June 12, 2019, Visa announced the launch of B2BConnect, a blockchain network of cross-border payment. On July 18, 2019, Facebook released a white paper of the Libra project, which aimed to build a simple, borderless currency system for non-sovereign countries based on the open source blockchain technology, and to build a financial infrastructure that could serve billions of people around the world. Meanwhile, governments have started to consider issuing their own central-bank digital currencies.

In 2014, the People's Bank of China started the research and development of the digital currency, namely Digital Currency Electronic Payment (DCEP), which was a new cryptocurrency system based on blockchain technology. Some typical projects in the world include Sweden's electronic krone, the Stella project jointly carried out by the Bank of Japan and the European Central Bank, the Jasper project of the Bank of Canada, the DNBcoin of the Dutch Central Bank, and the RSCoin of the UK.

2. Supply chain financing service

The *White Paper on Blockchain and Supply Chain Finance Service*, compiled by China Academy of Information and Communications Technology, Tencent Fintech and Shenzhen linklogis Financial Services Co., Ltd., pointed out that the supply chain financial solutions with blockchain as the bottom layer could release the credit of the core enterprise to the multi-level suppliers of the whole supply chain, improve the financing efficiency of the whole chain, reduce business costs, enrich the business scenarios of financial institutions, and thus improve the efficiency of capital operation in the whole supply chain.

In August 2019, Ant Financial cooperated with Chengdu Bank to transform traditional supply chain financing service with blockchain technology, and created the "double chain" mode, which had been run through first in Chengdu. Under this innovative mode, the bank could scan the information in the supply chain, knowing which large company's product line the order would eventually be supplied to, and that the loan would flow to the normal supply of the main business, and the borrower would have the ability to repay, so the loan risk would become more controllable.

In February 2020, the Beijing Municipal Government clearly proposed to build a blockchain-based credit and debt platform for supply chain to provide right confirmation and financing services for small, medium and micro enterprises participating in government procurement and state-owned enterprise procurement. Later, Beijing Financial Holding Group, together with Haidian District Government, Microchip Research Institute and other units, launched a blockchain based supply chain debt platform. According to People's Daily, on February 14, 2020, the platform completed the first online ownership affirmation and loan since its launch, helping a company obtaining a loan amounting to 720,000 yuan and the first loan of 440,000 yuan to support the fight against the Covid-19 epidemic and guarantee distance education in schools.

Blockchain can be a powerful tool for financial institutions in asset management (assets such as equity, bonds, bills, income vouchers, warehouse receipts, etc.) and user identity identification, etc. For example, the People's Bank of China has made some exploration in this aspect and applied blockchain technology to the bill trading platform. Blockchain will make use of its transparent and trusted features to achieve true financial inclusion.

5.3.2 Blockchain + Industry Innovation

1. Application in data privacy protection and data sharing in the medical industry

Most of the data in the medical industry involve personal privacy, which are extremely private. IBM Institute for Business Value (IBV) pointed out that blockchain technology would be of great value in the areas of clinical trial recording, regulatory compliance and medical/health monitoring records, also expertise in health management, data recording of medical device, drug therapy, billing and claims, safety of adverse event, medical asset management, medical contract management and so on. Healthcare providers, patients, and policy makers in the United States are looking for blockchain-based, portable and secure ways to digitally store medical records in order to create a one-stop medical record for an individual's lifetime. GEM, a blockchain technology provider, is working with a number of companies in the healthcare industry to develop a shared ledger and data security platform that is taking clinical cases from prototype to production.

2. Application in digital copyright protection

Blockchain technology publishes and reaches consensus on the ownership of digital rights with distributed ledger technology. Meanwhile, it ensures the uniqueness and unforgeability of copyright with timestamp technology, so as to solve the problems such as digital copyright registration, ownership affirmation, right protection. Many companies in the world have applied blockchain technology to digital copyright protection, including Monegraph, Colu, Binded, Singular DTV, and Chinese teams such as Ebookchain, Z-BaaS, and Yuanben Blockchain. Baidu and Beijing Qihoo Technology Co., Ltd. have actively launched project plan in the copyright protection of digital content based on blockchain.

3. Application in supply chain

In order to solve the problem of tracking the source of salmonella outbreaks in fruits, steaks and cakes, Walmart used blockchain to trace and track products. The efficiency has been greatly improved, and the tracing time for products has been reduced to 2 s. In May 2016, China Bulk Commodity Circulation Exchange Association of CFLP officially started the projects of "Online Account Registration Platform for Traders" and "Registration and Publishing Platform of the Electronic Warehouse Receipt", applying the blockchain technology to the logistics platform. Thanks to the "distributed ledgers" feature of the blockchain technology, the two obstacles of insufficient transparency in the transaction and inaccurate information in the storage and logistics in the bulk commodity trading have been solved.

4. Application in intelligent manufacturing

Philips and other global leading manufacturing enterprises have begun to launch plans in blockchain. Aircraft manufacturing giant, Airbus, announced to join the Hyperledger blockchain project, analyzing the source of suppliers and other components with the blockchain technology, and using the data of previous blocks to help Airbus reduce the time and cost of aircraft parts repair.

5. Application in clean energy

With the rapid development of clean energy, the vigorous promotion of "Internet plus" and the gradual maturity of intelligent grid technology, blockchain can realize digital and precise management of energy, which has great potential for restructuring energy transactions. LO3 Energy, an American energy company, has partnered with Consensus Systems, a bitcoin developer, to build Transactive Grid, an interactive power grid platform based on a blockchain system in communities in New York, such as BoerumHill, Park Slope, and Gowanus. Photovoltaic producers and electricity consumers on the platform do not depend on any power company. Transactions are made with each other with the blockchain system and settled in blockchain virtual currency, without third-party supervision.

5.3.3 Blockchain + Port and Shipping

In recent years, the world's major port and shipping enterprises have carried out the applications of blockchain technology in the areas of ship registration, document process optimization, cargo tracking, shipping insurance, etc. The applications of distributed ledger, smart contract, real-time consensus mechanism and other technologies in blockchain can improve the port and shipping supply chain network, help the shipping insurance industry, and promote the high-quality development of the traditional port and shipping industry.

1. Application in ship registration

The whole process of ship registration involves crossing borders and time zones, which is complicated and time-consuming. In May 2017, the Danish Maritime Authority launched the pilot project of Blockchain Ship Registration, which was also an important part of the Danish government's push for digital growth and aimed to maintain Denmark's position in the maritime sector. In September 2018, the Lloyd's Register of Shipping (LR) and Applied Blockchain in UK cooperated to establish a blockchain platform for ship registration, aiming to reduce the time needed, improve the efficiency, and explore extending the new technology to the shipping supply chain operations, and providing more values to other stakeholders.

2. Application in port and shipping EDI

Import and export shipping trades involve many entities such as sales, procurement, traders, carriers, port departments, ports and warehouses as well as series of supply chain information in trade, customs inspection and transportation organized around the bill of lading, including dozens of business documents. In order to realize paperless electronic data interchange and single window service between entities, freight forwarders, shipping companies, ports, customers and customs etc. use EDI technology among their respective application systems. Since various entities usually establish different data exchange formats and standards, the centralized service system results in non-uniform shipping data standards and complicated data exchange formats, which exposes many problems in data security and information leakage.

The decentralization of blockchain can remove the core position of traditional EDI center, and relevant entities in the supply chain become equal nodes in the P2P data network. The traceability of blockchain enables shippers, customs, carriers and insurance companies to trace reliable electronic evidence, clearly define the responsibilities of all parties, and improve the processing efficiency of payment, delivery and claim settlement. In 2015, an Israeli startup, Wave, began to use blockchain technology to solve the problem of inconsistent EDI data standards and lack of trust. All payments or documents must be agreed by all parties, thus establishing information trust, eliminating all concerns in security and data, and improving efficiency. In early November, 2016, the world's first blockchain logistics research alliance in the field of logistics was jointly established by the Rotterdam Port, the Netherlands National Institute of Applied Sciences, Algemene Bank Nederland, Delft University of Technology, Flower Trading Center of Desheim University of Applied Sciences, to optimize the data management of rose transportation from Kenya in Africa to Rotterdam in the Netherlands by using blockchain technology, transferring the document processing related to container shipping to the smart contract system, and promoting the document process optimization.

3.   Application in shipping network optimization

80% of goods in global trade are transported through ports by sea. The immutability and distributed storage of blockchain technology can help build a highly secure and transparent shared network of shipping supply chains. Each participant sees the progress of goods in the supply chain according to the authority level and knows where the container is going. All participants in supply chain management can exchange shipping information in real time, safely and seamlessly.

In 2018, Maersk Line and IBM started cooperation on blockchain technology application, jointly developed the blockchain transportation solution, TradeLens, by establishing a single shared view of transactions that protects the details, privacy or confidentiality of transactions. TradeLens members could access to data and documents of their global cargos in real time in the transportation process, including the IOT and sensor data (such as temperature control, container weight, etc.), so that the shippers, shipping companies, freight forwarders, port and terminal operators, inland transportation carriers and customs could more effectively interchange data. By the end of 2019, TradeLens had got more than 100 participants on the platform, and recorded more than 10 million shipping events as well as handled tens of thousands of documents per week.

In 2018, Dubai PSA World, Hutchison Port Group, PSA International Port Services Group, Shanghai International Port Group, France CGM Group, COSCO Shipping and Container Lines, Evergreen Marine, OOCL, Yang Ming Shipping and software solution provider CargoSmart signed a letter of intent, and reached cooperation intention on building the shipping blockchain alliance Global Shipping Business Network (GSBN). In December 2019, the Shanghai International Port Group, COSCO Shipping and Container Lines realized data flow and mutual trust in cooperative processes between the shipping company system, the terminal system and the blockchain platform with GSBN. Customers could complete the operation process on the chain at one time throughout the shipping company and the port, to realize the paperless import shipment release in the whole process and ensure zero delay of customers' import business.

4.   Application in shipping finance and insurance service

In 2018, Block Shipping in Denmark launched the Global Shared Container Platform (GSCP), which realized transaction settlement and clearing in the form of blockchain cryptocurrency.

Since shipping insurance is often involved in transnational business, with numerous various documents and copies, and transactions of large volume, the current operation process of shipping insurance is cumbersome and tedious, and difficult to check accounts. In 2018, Ernst & Young, an international accounting and consulting firm, partnered with Guardtime to launch the world's first shipping insurance blockchain platform for commercial use. Global shipping giants Maersk, Microsoft, American Insurance Standards Association, MS Amlin Insurance, Truly Insurance, etc., jointly participated in the establishment of the platform. Parties in the shipping insurance ecosystem use distributed ledger technology to record shipping information and automate insurance transactions.

The global blockchain development still has technical difficulties in consensus mechanism, distributed storage, database, security and other aspects. For instance, the current consensus mechanism of blockchain still has the problems of excessive loss, high cost, few types, and difficult to match diversified application scenarios. The small block capacity leads to the congestion of the current blockchain network, and it is still difficult to meet the ideal high-intensity and high-frequency business requirements. Blockchain technology, as one of the most popular technologies at present, needs more time and energy to be invested in its learning and research.

Bibliography

1. Swan M (2015) Blockchain: blueprint for a new economy. O'Reilly Media, Sebastopol
2. Pilkington M (2016) Blockchain technology: principles and applications. In: Research handbook on digital transformations. Edward Elgar Publishing
3. Nofer M, Gomber P, Hinz O et al (2017) Blockchain. Bus Inf Syst Eng 59(3):183–187
4. Zheng Z, Xie S, Dai HN et al (2018) Blockchain challenges and opportunities: a survey. Int J Web Grid Serv 14(4):352–375
5. Wang J, Ye L, Gao RX et al (2019) Digital Twin for rotating machinery fault diagnosis in smart manufacturing. Int J Prod Res 57(12):3920–3934
6. Xu Y, Sun Y, Liu X et al (2019) A digital-twin-assisted fault diagnosis using deep transfer learning. IEEE Access 7:19990–19999

Chapter 6
Smart Port and Artificial Intelligence

Artificial intelligence (AI), as a newly developed subject, was put forward decades ago. Computing power resources, big data and emerging AI methods are the main driving forces for the vigorous development of AI in the past decade. Dedicated weak AI has made great progress and development, and its performance in some fields has reached or even exceeded the human level. However, there are still many fields and scenarios in which AI research is still at its initial stage. Therefore, for a long time in the future, the research and development of AI in the professional field will still be the focus of AI research. The study of strong AI just begins. As far as the current researches are concern, despite many positive research attempts, strong AI still has a long way to go without more ground-breaking approaches. Even so, every weak AI study is laying the groundwork for the strong AI. In the port field, some relevant enterprises have applied AI technology in some of their core business scenarios, including intelligent container collection, intelligent stowage, intelligent ship control, etc., and achieved certain results. There is no doubt that further investment in the research and development of various special AI in the port field is a necessary way to construct the smart port.

6.1 General Introduction of AI

6.1.1 Concept of AI

In the summer of 1956, John McCarthy and other scholars held a meeting in Dartmouth College in the United States to discuss "how to use machines to simulate human intelligence". The meeting put forward the concept of "artificial intelligence", marking the birth of AI. There have been various definitions in the development of AI up to now, which mainly include the following four dimensions.

From Fig. 6.1, it can be seen that the upper two dimensions of the quadrium focus on thinking, the lower two on action, the left two on the fidelity to human, and the

© Shanghai Scientific and Technical Publishers and Springer Nature Singapore Pte Ltd. 2022
W. Mi and Y. Liu, *Smart Ports*, https://doi.org/10.1007/978-981-16-9889-7_6

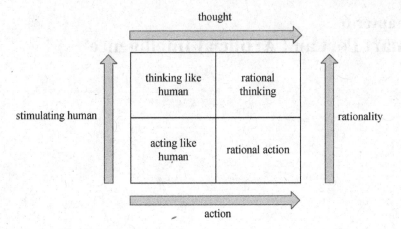

Fig. 6.1   Four dimensions of AI

right two on rationality. At the beginning, researchers paid more attention to how to mimic the working of human mind, but as the research went on, more scholars believed that it was more important to study the basic principles of intelligence than to copy. For example, the dream of "human flight" took off when people stopped trying to imitate birds and began to understand aerodynamics and use wind tunnels. No aeronautical engineering textbook would define the goal of its field as being able to fly exactly like a bird in order to fool other real birds. Therefore, the research on AI has gradually shifted from simulating the working mechanism of the human mind to expanding its basic principles to make more rational thinking, decision-making and behavior.

To sum up, the current definition of AI is more generally accepted as: AI is a new technology and science that studies and develops theories, methods, technologies and application systems used to simulate, extend and expand human intelligence.

6.1.2   Fields of AI

At present, the research on AI includes three fields: intelligent perception, intelligent decision and intelligent control. These three supplement each other, consisting of large-scale AI systems.

1.   Intelligent perception

Intelligent perception refers to the ability to obtain and understand information with a variety of sensors, which has two aspects. The first is information acquisition. Human beings have rich sensory organs, which can feel external stimuli to obtain environmental information through vision, hearing, taste, touch and smell. Mechanical devices can also obtain environmental information around them with a variety of sensors. The second is information fusion and understanding, which is to effectively fuse the information acquired by different sensors and understand and infer the fused information with the intelligent algorithms.

2. Intelligent decision

Intelligent decision is to make clear the goal and constraint conditions for the decision problem, and make reasonable decision quickly for the complex planning and scheduling problems. Intelligent decision-making consists of three levels. The first is knowledge representation and storage, which extracts historical data and abstracts existing experience to synthesize "knowledge" for storage. The second is planning and decision-making, which uses the stored knowledge of intelligent algorithm to plan and make decisions. The last is the learning ability, which evaluates the decision results, and extracts new information and knowledge to continuously optimize the decision agent.

3. Intelligent control

Intelligent control means that the equipment can identify and determine the operating target and purpose independently, and complete the operation task accurately, safely, efficiently and automatically. Usually, intelligent control needs to be combined with intelligent perception, using the external information obtained by intelligent perception and its understanding results to make decisions and control the equipment to make reasonable actions.

6.1.3 Categorization of AI

1. Weak AI

Weak AI, also known as Narrow AI or Applied AI, refers to AI that can only accomplish a specific task or solve a specific problem. Weak AI does not mean that it is weak in problem-solving, but that it is weak in versatility and can only be used to solve problems in a specific field. Alpha Go, for example, can master the Go game but can't play gobang.

At present, main scientific research focuses on weak AI, and it is generally believed that considerable achievements have been made in this field.

2. Strong AI

Strong AI, also known as Artificial General Intelligence (AGI) or Full AI, refers to intelligent machines that can perform any intellectual task in the same way as humans. Such AI is the ultimate goal in part of the AI research, and has been a persistent topic in many science fictions.

There is no precise definition of the intelligence level required for strong AI, but AI researchers agree that strong AI requires the following:

①   Thinking ability, using strategies to solve problems, and making judgments in uncertain situations;
②   Showing a certain amount of knowledge
③   Planning ability
④   Learning ability
⑤   Communicating ability
⑥   The ability to use all of one's abilities to achieve a goal.

Compared with weak AI, the research of strong AI is in a stagnant condition.

3.   Artificial Super Intelligence (ASI)

ASI is a kind of super intelligence. Nick Bostrom, a philosopher and director of the Institute for the Future of Humanity at Oxford University, defined super intelligence as "any intelligence that substantially exceeds human cognitive performance in almost all areas". First of all, ASI can achieve the same function as human intelligence, that is, it can reprogram and improve itself in the same way that human intelligence has evolved biologically, which is called "recursive self-improvement function". Second, Bostrom mentioned that "a biological neuron fires at 200 Hz, but even a present-day transistor operates at the Gigahertz" and that "neurons propagate slowly in axons, but in computers, signals can travel at the speed of light." This will make the ASI think faster and self-improve faster than humans, and all the biological limitations of humans will be broken by machine intelligence.

6.1.4 Methods in AI

In the development of AI, researchers have put forward many AI methods for different scenarios and problems, different methods also have different scopes of application, among which the more classic and popular methods include but are not limited to the following categories.

1.   Heuristic algorithm

Heuristic algorithm is a kind of algorithm based on intuitionistic or experiential construction. Compared with optimization algorithm, heuristic algorithm is usually used to solve complex combinatorial optimization problems. Most complex decision-making problems encountered in real life are problems that can be solved in nondeterministic polynomial time, usually known as NP-hard or NP-complete problems. Therefore, heuristic algorithm is usually used to solve the effective approximate solution of this kind of problem.

Common heuristic algorithms include: simulated annealing (SA) algorithm, genetic algorithm (GA), ant colony optimization (ACO), etc. These algorithms are "inspired" by some biological behaviors or some phenomena and laws in the process of life evolution, so they can also be regarded as the algorithms to simulate human or biological intelligence.

2.   Deep learning

Deep learning belongs to the category of machine learning algorithms, which is usually suitable for solving pattern recognition problems. The concept is derived from neural networks, and the multi-layered perceptron with multiple hidden layers is a deep learning structure. Deep learning combines features of lower layers to form more abstract high layers representing attribute categories or features to discover distributed characteristic representations of data. The motivation for deep learning is to build neural networks that mimic the human brain for analytical learning, and to interpret data, such as images, sounds and texts, by mimicking the mechanisms of human brain.

The popularity of deep learning methods even leads to the misunderstanding that AI equals deep learning. Deep learning can be said to be the leader of AI methods at the present stage. However, with the deepening of research, the shortcomings of deep learning methods are gradually revealed, such as the poor interpretability of its black box model and the failure of the model easily caused by noise interference. Despite the flaws, deep learning remains by far one of the most important and effective methods in AI.

3.   Reinforcement learning

Reinforcement learning also belongs to the category of machine learning algorithms, which is usually used for machine control, sequential decision-making and other problems. Reinforcement learning refers to the rewarding behavior that agents acquire through interaction with the environment. The goal of reinforcement learning is to make agents obtain the maximum cumulative reward return. Reinforcement learning is different from supervised learning in that it is mainly manifested in reinforcement signals. In reinforcement learning, reinforcement signals provided by the environment are used to evaluate the quality of the generated actions (usually scalar signals), rather than to tell the reinforcement learning system how to produce the correct actions. Since the external environment provides little information, agents must learn from their own experiences. In this way, agents acquire knowledge in an action-evaluation environment and optimize action plans to suit the environment. Therefore, reinforcement learning can be called an algorithm to simulate human learning behavior. Meanwhile, reinforcement learning is also endowed with more powerful learning ability with the combination of deep learning method. The famous Go AI AlphaGo Zero gradually learned and mastered the game of Go by playing self-game through reinforcement learning.

6.2   Development of AI

1.   In intelligent perception

In 2010, Professor Li Feifei launched the first ImageNet Large-Scale Image Recognition Challenge based on the image database he created. From 2010 to 2017, seven consecutive events were held, and the annual recognition accuracy rate increased to an astonishing 97.75% from the original 71.8%. At the end of 2017, the ImageNet founder Li Feifei announced that the contest would no longer be held. The reason was that in the image recognition of ordinary scene, the accuracy rate of AI recognition was much higher than that of human beings, and the algorithm had little room for improvement. It made little sense to continue to compete in depth, and brought the curtain down on the ImageNet contest.

A new generation of AI methods is also making breakthroughs in speech recognition. In March 2017, IBM combined the LSTM model with the WaveNet language model, which had three strong acoustic models. This reduced the error rate to 5.5%, slightly lower than the error rate of professional stenographers (5.1%). In August 2017, Microsoft reduced the error rate to 5.1%, the standard for professional human stenographers, by improving the neural network-based auditory and speech models in its speech recognition system. In June 2018, Alibaba's Damo launched a new generation of speech recognition model, DFSMN, which improved the world record of speech recognition accuracy to 96.04% and reduced the error rate to 3.96%. In October 2018, Cloudwalk released the new Pyramidal-FSMN speech recognition model, which reduced the error rate to 2.97%, much lower than the professional stenographer.

2.   In intelligent decision-making

In 2015, Google's Deep Mind team put forward the concept of deep reinforcement learning and Deep Q Learning algorithm for the first time. With pure images as input, the algorithm used a combination of reinforcement learning and deep neural networks to train the AI system to master nearly 2,600 playing techniques of Atari Games and surpassed the level of top human players, which proved that the AI system could discover the internal strategy of simple control tasks by autonomous learning, and continuously manipulate its own behavior to achieve the possibility of performing these control tasks beyond the human level.

In 2016, based on the games of human Go masters, Google's Deep Mind team trained the AI system AlphaGo in the way of deep learning combined with Monte Carlo tree search, which beat Lee Se-dol by a score of 4 to 1, further proving that the AI systems could learn from the experience of experts in the field to discover and refine potential complex strategies.

AlphaGo still drew on the experience of human Go masters, while AlphaGo Zero, which was launched in 2017, was an intelligent body learning and evolved by playing Go games against itself. It outplayed AlphaGo in three days by playing against itself, winning 100 matches and reached the level of AlphaGo Master in 21 days and surpassed all previous versions in 40 days. Its ability to refine complex strategies entirely through autonomous learning, without relying on any human experience, was a big step forward compared to traditional AI technologies.

In 2019, Alpha Star's AI challenged top human players in real-time strategy (RTS) game. The decision complexity of RTS game was much higher than that of Go game, and the decision space was about one million times as large as that of Go game, but in the end, the AI still defeated human beings by an overwhelming result of 5:0.

All the above results have shown that AI has immeasurable potential in the field of intelligent decision-making.

3. In intelligent control

Intelligent control robot is widely used in the field of industrial manufacturing because of its high accuracy and high efficiency in performing tasks. In the automobile production industry, assembly of parts and other tasks all need the assistance of intelligent robots. As it is a metronomic assembly line, the actions and time of each station can be strictly controlled, the intelligent robots can complete a large number of mechanical assembly work in a short time. Compared with the robots used in the field of industrial manufacturing mentioned above, there is another kind of intelligent control robot that is hot in the current research, that is, the robot that can perform uncertain actions. Uncertainty refers to the actions of interacting objects in the environment, such as opening a door or opening a bottle, which is very different from the automobile manufacturing assembly line. Compared with the fixed assembly line, the position of the door, the shape of the door handle, the shape of the bottle cap and the opening way of the bottle cap may be different. Traditional control methods can accurately model a particular door, but the robot will not be flexible once there are more kinds of doors. Aiming at such intelligent control problems, the mainstream approach now is to make use of the powerful generalization ability of deep learning model to make the robot adapt its posture and action according to the environment, and some achievements have been made. However, the research in this field is still at the initial stage, and the stability of the intelligent control of uncertain motion needs to be further improved (Fig. 6.2).

To sum up, the current research on ANI has made an important breakthrough. Because of the features of single task, clear demand, clear application boundary, rich domain knowledge and relatively simple modeling, the ANI system for specific tasks has made a single breakthrough in the field of AI, and can surpass human intelligence in the single test of local intelligence level. The recent progress of AI is mainly concentrated in the field of ANI, while the research progress of AGI is slow,

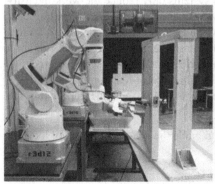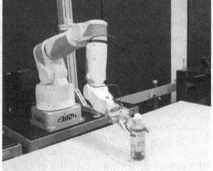

Fig. 6.2 Intelligent control of uncertain motion

there is no transformative method, and it is difficult to make a breakthrough in the short term. Therefore, in the future for a long time, the research of AI is bound to continue to focus on various professional fields, and each innovation of ANI is to contribute to the journey towards AGI and ASI. As AI scientist Aaron Saenz put it, "our world's ANI systems are like the amino acids in the early Earth's primordial ooze, the inanimate stuff of life that, one unexpected day, woke up."

6.3 Applications of AI in Smart Port

In recent years, under the background of Intelligent Manufacturing 2025, a wave of construction of automated terminals has been initiated in China. Meanwhile, the trend of intelligent transformation of the traditional manual terminal is becoming more and more intense, which has put forward new and higher requirements for the daily operation planning and operation scheduling. AI-related technologies and methods have been widely applied in all aspects of port operations. This section will focus on the applications of the intelligent decision-making in the port. Examples of applications of the intelligent perception will be discussed separately in Chap. 7.

6.3.1 Intelligent Container Collection

The condition of container stacking in the terminal yard has a direct impact on the efficiency of the four operations of loading, unloading, collection and delivery in the terminal. How to select the appropriate location for container is the most important issue to improve the efficiency of terminal operations. Therefore, the intelligent location selection of export containers in container terminal is the most critical technical part in the intelligent transformation process of traditional manual terminals.

At present, the intelligent location selection of container terminal is still at the initial stage. Most of the location selection systems only reduce the work intensity of the staff, and have little improvement on the operation efficiency of the terminal. The intelligent location selection system of the next generation container terminal should guide the operation, taking improving the efficiency of the terminal operation as the main goal, and the rational use of historical data and stowing requirements as the key factors to achieve this goal. The significance of realizing the data-driven intelligent location selection system of the next generation container terminal lies in the following.

1. Extracting the value of historical data to guide the decision-making of location selection

In the era of data explosion, the intelligent location selection system of container terminal can make full use of the historical data accumulated over years, and analyze and extract the massive historical data to mine its potential objective law as the priori knowledge, so as to provide a more reliable and detailed reference for the intelligent location selection decision.

2.   Pull strategy of container collection for the purpose of stowage

Combined with the intelligent stowage system, the intelligent location selection system is pulled to make decisions based on the stacking requirements of the stowage plan, which serves for the stowage plan and shipping delivery. The ultimate purpose of the intelligent location selection of export containers is to cooperate with the stowage plan and the actual shipping delivery operation. Therefore, the stowage requirement is of great significance to guide the intelligent location selection decision. It is also an important factor to make full use of the experience accumulated by the intelligent stowage system to provide support for the fine-grained location selection decision.

3.   Dynamic planning mode to improve the utilization of yard resources

A major advantage of the intelligent location selection system of the next generation container terminals is that there is no need for manual yard planning, and all location selection decisions are made dynamically by the system at the moment when the container enters the terminal through the gate. These decisions are based on historical data as a reference, which can be used to calculate the location selection rules more finely without the support of yard planning. Meanwhile, it can avoid the phenomenon that yard planning occupies yard resources in advance, so as to improve the utilization rate of yard resources.

4.   Subdividing operation technique modes

According to different ship loading process modes of different voyages, specific location selection rules are formulated. In recent years, the continuous emergence of VLVs has brought more challenges to terminal operations. In order to better serve VLVs, terminal operators have put forward many new handling technique plans, such as concurrent loading and unloading, reverse stowage in the cabin and centralized stacking of custom released containers, etc., which have put forward new requirements for container stacking. By subdividing the operation technique modes, the configurable strategy of container collection and location selection is realized to make the result more in line with the actual requirements of loading and indirectly improve the efficiency of loading operation.

5.   Reducing the work intensity of staff

Another major advantage of intelligent location selection is that it has eliminated the work of manual yard planning, port and tonnage grouping, and assigning locations for containers without yard plan, which is limited by experience and the working condition of staff. The experience will be solidified with intelligent computing, and the computer will automatically take over the relevant work. The terminal staff can be liberated from the trivial and complicated work, and have more time to find problems, summarize the rules, and further propose the optimization plan for the system, so as to help the intelligent system to further improve the location selection results.

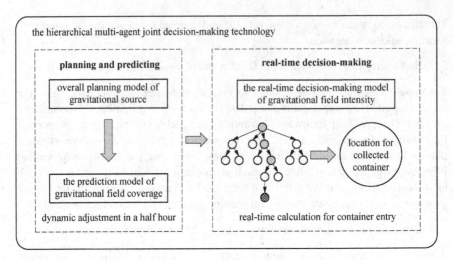

Fig. 6.3 The overall structure of the hierarchical multi-agent joint decision-making technology with the gravitational field mechanism for container collection

The intelligent container collection adopts the hierarchical multi-agent joint decision-making technology with the gravitational field mechanism as the core, including the overall planning of gravitational source, the prediction of gravitational field coverage, and the real-time decision-making of gravitational field intensity. The overall structure is shown in Fig. 6.3.

As can be seen from Fig. 6.3, intelligent container collection is a comprehensive intelligent system which combines planning, prediction and real-time decision-making. Among them, the gravitational field mechanism is the core mechanism of the intelligent location selection algorithm, which determines the degree of gravitation of the yard area to any coming target container. The rationality of the container collection decision can be realized through the reasonable planning of the gravitational field area for all the locations in the yard. The gravitational field mechanism mainly consists of the following three core intelligent modules.

(1)   Overall planning model of gravitational source

We regard containers of different ships, different voyages, different container groups and different tonnage levels as independent gravitational sources. The purpose of the overall planning of gravitational source is to ensure that the containers of the same kind can attract each other and form a centralized stacking, while containers of different kinds need to be mutually exclusive to avoid conflict in subsequent loading operations.

(2)   Prediction model of gravitational field coverage

When the location of the gravitational source is determined, the next step is to determine the coverage of each gravitational field. The intelligent container collection system combines historical data, actual container entry data and pre-input container data to build a deep neural network to predict the coverage of the gravitational field, so as to determine the reasonable coverage of each gravitational field. The accuracy of the prediction of the coverage of the gravitational field will have a great impact on the utilization rate of the yard resources. Too large predicted coverage will result in waste of spare yard resources; otherwise, it will result in too scattered stacking, unable to form a centralized stacking situation, which will affect the efficiency of subsequent loading operations.

(3) Real-time decision-making model of gravitational field intensity

When the location of the gravitational source and the coverage of the gravitational field are determined, it is necessary to wait for the arrival of the container to make the final selection decision. Since the general location of container collection has been determined, the specific decision of container collection location is more focused on considering the use of resources such as facility resources in the real-time container block and the busyness of the yard. By combining the usage of these real-time resources and the planned gravitational field information, the intelligent container collection system dynamically constructs a Monte Carlo decision tree to determine the most appropriate container collection location for the current coming container.

6.3.2 Intelligent Stowage

The stowage plan of container ship is a key part in the loading plan, the optimization level of the decision largely determines the efficiency and cost of terminal operation. At present, the bottleneck of manual stowage decision mainly includes the following.

(1) The stowage efficiency is low. When the loading volume exceeds 1,500 containers, it may take more than 3 h to make a stowage decision.
(2) The global optimization of decision-making is unable to achieve and it is difficult to calculate the global optimization level with manual decision.
(3) The stability of decision-making is relatively low, and it completely depends on the personal experience of the stowage operator. Individual differences may lead to unstable stowage results.
(4) The cost of personnel training is high, and the stowage personnel often need more than half a year's training before they can start working.

The working conditions in container terminal stowage decision-making process are ever-changing. Experienced stowage operators can accurately judge the current stowage conditions according to the pre-stowage biplane and container distribution in the yard, but these working conditions are difficult to accurately describe with words. The stowage strategies adopted by the stowage operators under different working conditions also cannot be described with simple rules. Therefore, it is necessary to construct a deep learning model to abstract the stowage condition and stowage strategy respectively, so as to solve the problem of accurate extraction and storage of stowage knowledge and complex work conditions. Among them, the stowage condition network converts the historical stowage data of different ships at each bay into a sparse matrix of constant size and builds a deep neural network for the recognition of stowage conditions. The stowage condition network is shown in Fig. 6.4.

The design of the stowage strategy network is to take the summarized basic stowage principles as basic inputs and abstract them to form more complex stowage strategies. Therefore, the historical stowage steps and the stowage reference are taken as the input to establish a fully connected restricted Boltzmann machine in the form of cascaded autoencoders, thus constructing the stowage strategy network (Fig. 6.5).

The results of training through these two networks provide the core decision basis for the realization of intelligent stowage. Meanwhile, when the existing knowledge is unable to make effective decisions for the current ship, the two stowage networks can be further improved by manually adjusting the stowage results and using the

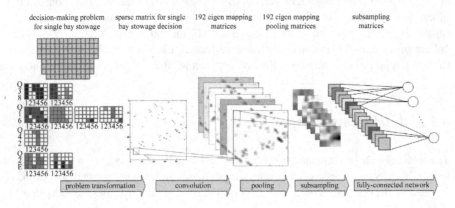

Fig. 6.4 Stowage condition network

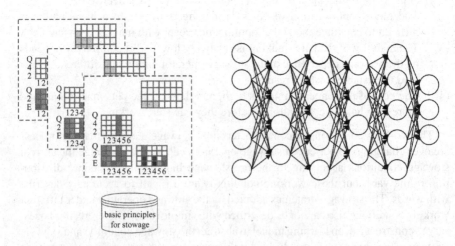

Fig. 6.5 Stowage strategy network

adjusted results to enhance the training of the network, so as to realize the evolution of the intelligent stowage system.

A rule scheduling method has been put forward based on workflow engine, which provides a reasonable knowledge network scheduling plan for intelligent stowage decision-making under complex conditions, overcoming the problem of low efficiency caused the exponential growth of solution space in large-scale decision-making process and, solving the key problem of pruning strategy organization under different working conditions in the process of intelligent stowage. The engine can prune the current decision space by pattern matching for the current decision condition, identifying and scheduling different rules. For some non-simple conditions, the engine can select and reconstruct the knowledge stack suitable for the current decision-making condition from the knowledge network, optimize the state nodes to be searched, and reduce the decision-making space significantly.

The traditional methods to solve the stowage problem are mainly intelligent algorithm (such as genetic algorithm, tabu search and hybrid algorithm) and heuristic algorithm. The required search space of stowage problem is very large, the constraints are very complex, and the constraints are not the same under different working conditions, therefore, the traditional solution method has the following limitations.

(1) The constraint rules are too complex, and different algorithms or different heuristic methods need to be designed for different working conditions, resulting in low reusability.

(2) The required solution space is too large, the convergence efficiency is low, the solution time is too long, and it is difficult to get a satisfactory solution.

Therefore, an efficient method to organize decision-making rules is needed, and the search space can be reduced by pruning to improve the efficiency and effect of solution.

The scheduling method is designed based on workflow engine to effectively deal with the correlation between operation process and solution flow. In the solution process, according to the features of the current solution node, through the workflow engine's historical data analysis and process analysis, the corresponding working conditions are matched, the constraint set of the current node is obtained, the current sub-search node is generated, and the required solution strategy is identified. If there is no working condition matching in the historical database, it means that the current node is a new working condition. Meanwhile, the workflow engine obtains the corresponding solution strategy based on the analysis of the features of the new condition and the operation process. After the extraction of solving strategies is completed, the knowledge stack is reconstructed according to the priority of strategies to form a strategy set for the current node and search for each child node.

After the solution is completed, the strategy and pruning effect are analyzed in reverse according to the optimal path finally selected for the constructed solution tree. By calculating the difference between the actual expected value and the actual value of the sub-nodes in each solving step, the residuals are learned, the corresponding parameters of the strategy network are corrected, and the revised knowledge is explained through workflow analysis. Then the condition features and strategy features of the newly added conditions in the solution process are updated, so that the next time when the same condition is encountered, it can be solved by matching features directly. Operation process of workflow engine is shown in Fig. 6.6.

The application of multi-layered rule screening has effectively solved the problems of poor fitting degree and low solving efficiency in the solution method of stowage problem when facing complex multi-classification problems, and made the optimization results reach or even exceed manual stowage, reducing the decision-making time of stowing thousands of containers from about 3 h in manual stowage to 5 min.

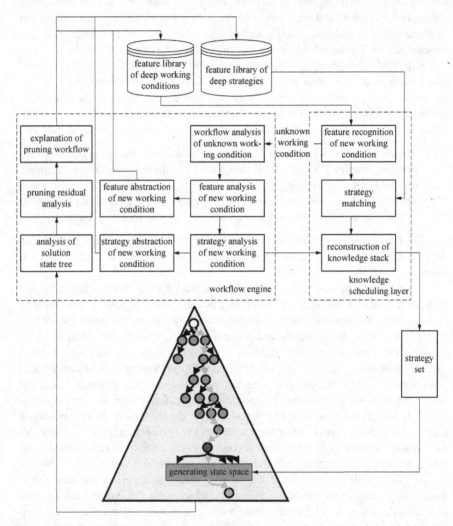

Fig. 6.6 Operation process of workflow engine

6.3.3 Intelligent Ship Control

Ship loading and unloading operation is the most important operations of a container terminal. The control of ship loading and unloading operation instruction sequence has a direct impact on the overall operation efficiency of a terminal. How to reasonably control the instruction sequence of ship loading and unloading operation is the most important issue to improve the operation efficiency of a terminal. Therefore, it is one of the most critical technical steps in the intelligent transformation of traditional manual container terminal to realize the intelligent control of command sequence in ship loading and unloading.

At present, there are mainly the following problems in control of ship loading instruction sequence in the container terminals.

(1)   The control of instruction sequence from the scheduling operators in the central control office depends too much on human experience, and staff of different levels have different control degrees on the rationality of the instructions.

(2)   Different scheduling operators in the central control office make decisions on the loading sequences of different ships, and there is no communication between them, which may easily lead to unreasonable facility scheduling and thus reduce the efficiency of ship loading and unloading.

(3)   The lack of balance mechanism of operation efficiency of all the operation lines along the shoreline is liable to cause the problem of conflict between operation instructions.

(4)   Position selection of unloaded containers depends on manual planning, which has poor dynamic property, and often causes the situation of containers without yard plan under high operating pressure.

The current instruction sequence control mode relies too much on human experience, and the scheduling operators lack effective means of communication and sufficient time to reasonably think about the allocation of relevant resources, which leads to the failure in maintaining a high level of the overall operation efficiency of the terminal. While the intelligent ship control system aims to build an intelligent control system of loading and unloading instruction sequence along the whole shoreline from a global perspective, according to the current operation situation, the efficiency of each operation line is balanced in real time, so as to improve the overall efficiency of loading and unloading operation of the terminal. The significance of realizing the intelligent control of the instruction sequence of ship loading and unloading along the whole shoreline lies in the following.

(1) Real-time dynamic control of the instruction sequence along the whole shoreline rationally arranges the container delivery points and unloading positions of each operation line to ensure the operation efficiency of ship loading and unloading of the whole terminal. A big advantage of the next generation of intelligent control system of ship loading instruction sequence in the container terminal is to take over the decision-making of the container delivery and unloading instruction control of all the ships currently under operation along the terminal shoreline, and compute

the container delivery points in the yard for each operation line from a more global perspective, thus to ensure the overall efficiency of the ship loading operation.

(2) The computer can make specific instruction control rules according to different voyages and different loading technique modes. In recent years, the continuous emergence of ultra-large ships has brought more challenges to terminal operations. In order to better serve the ultra-large ships, operators in the terminal have also put forward many new handling techniques, such as concurrent loading and unloading. And these new handling techniques all have raised new requirements for the instruction control of loading operations. By refining the operation modes, the configurable instruction control rules are implemented to ensure the efficiency of loading operations in the container terminal under different operation modes.

(3) It is capable of automatic control of container delivery points of the whole yard, automatic identification of the current key operation lines, automatic and dynamic instruction control of unloading containers as well as position selection and automatic balance of the operation efficiency along the shoreline, reducing the work intensity of staff. Another major advantage of intelligent control of instruction sequence is that it eliminates a lot of complicated work which is limited by human experience and the working condition of operators. It solidifies the human experience by means of intelligent computing and the computer automatically takes over the related work. While the operators can be liberated from the trivial and complicated work, and have more time to find problems, summarize the rules, and further put forward the plan to optimize the system, so as to help optimize the decision-making results of the intelligent system.

The intelligent ship control adopts the joint decision-making technology of multi-layered agents to form a two-layered decision-making structure of dynamic short-term planning of container delivery operation instruction and dynamic decision-making of container delivery operation instruction. At each layer, fusion calculation of multi-agents is implemented to achieve quick solution on the basis of decomposing large-scale complex planning problems. Dynamic short-term planning is to classify and aggregate the operating containers within a single bay of each operation line according to the current task execution and yard facility conditions every period of time, decomposing the operations to obtain the short-term scope of the current operation. Subsequent instruction scheduling is carried out within the scope of short-term plan to improve the computational efficiency of instruction decision. According to the operation scope provided by the short-term plan, the specific instruction sequence activation is decided by decision-making system of dynamic container delivery instruction in line with the operating progress requirements of the current operation line. According to the logical structure of hierarchical multi-agent joint decision-making, the core of intelligent ship control is the two models below.

1. Dynamic short-term planning model

The short-term dynamic planning model calculates to obtain the loading sequence of the blocks currently waiting for loading operation with the computing network of the main blocks based on the current operation task distribution in the yard and facility condition and location, in which the current main blocks and auxiliary blocks can be determined to avoid the conflict between operation lines, to optimize the operation efficiency of all the operation lines in the premise of reducing mutual interference between the operation lines. Then, according to the distribution of containers to be delivered in the blocks, the specific container delivery bays and the number of short plans in the main and auxiliary blocks are determined by the calculation of container delivery point network, and the current dynamic short plans are obtained.

2. Dynamic instruction decision-making model

The dynamic instruction decision-making model obtains the transition tree structure of the state space under the current short-term plans according to the real-time situation of the current facilities and instructions. Dynamic instruction network combined with Monte Carlo tree search algorithm is applied in the short-term planning transition tree structure of the state space, considering the constraints including safety requirements of operating on deck and in cabin, yard facility and yard traffic conditions, etc., to optimize container reshuffle, facility shifting and quay crane utilization, and determine the final dynamic instruction sequence in order to improve the operation efficiency and reduce the operation cost.

Through layered multi-agent joint ship control decision-making, efficient and stable intelligent ship loading scheduling is realized, which has effectively alleviated the operation pressure of scheduling operators in the process of loading and unloading operations in large container terminals, further improved the loading and unloading efficiency of container terminals, and reduced the operation cost.

Bibliography

1. Searle JR (1980) Minds, brains, and programs. Behav Brain Sci 3(3):417–424
2. Crevier D (1993) AI: the tumultuous history of the search for artificial intelligence. Basic Books, NY
3. McCorduck P (2004) Machines who think: a personal inquiry into the history and prospects of artificial intelligence. CRC Press, Florida
4. Russel S, Norvig P (2013) Artificial intelligence: a modern approach. University of California, Berkeley
5. Brundage M (2015) Taking superintelligence seriously: superintelligence: paths, dangers, strategies by Nick Bostrom (Oxford University Press, 2014). Futures 72:32–35
6. Mnih V, Kavukcuoglu K, Silver D et al (2015) Human-level control through deep reinforcement learning. Nature 518(7540):529
7. Silver D, Huang A, Maddison CJ et al (2016) Mastering the game of go with deep neural network and tree search. Nature 529(7587):484–489
8. Zhao N, Shen Y, Xia M et al (2016) A novel strategy for stowage planning of 40 feet containers in container terminals. J Mar Sci Technol 24(1):9
9. Shen Y, Zhao N, Xia M et al (2017) A deep q-learning network for ship stowage planning problem. Polish Maritime Res
10. Silver D, Schrittwieser J, Simonyan K et al (2017) Mastering the game of go without human knowledge. Nature 550(7676):354–359

11. Zhao N, Guo Y, Xiang T et al (2018) Container ship stowage based on Monte Carlo tree search. J Coast Res 83(10083):540–547
12. Vinyals O, Babuschkin I, Chung J et al (2019) Alphastar: mastering the real-time strategy game starcraft II. DeepMind Blog 2
13. Xia M, Li Y, Shen Y et al (2020) Loading sequencing problem in container terminal with deep Q-learning. J Coast Res 103(SI):817–821

Chapter 7
Smart Port and Machine Vision

In modern industrial production systems, there are a lot of part inspection, identification and process monitoring. Traditionally, such highly repetitive processes require a large number of labor hours and are not very efficient. With the rapid development of computer technology and sensor technology at the end of the twentieth century, people began to try to use computers to replace manual image processing and recognition, which led to the birth of machine vision technology. As defined by the Machine Vision Section of the Society of Manufacturing Engineers (SME) and the Robotic Industries Association (RIA), "machine vision is the use of devices for optical non-contact sensing to automatically receive and interpret an image of a real scene in order to obtain information and/or control machines or processes."

7.1 General Introduction to Machine Vision

After many years of development, machine vision technology has been widely applied, forming a specific field of expertise, while its concept and application fields are also extended. Now, machine vision technology is not limited to the traditional image recognition, it is also widely used in image enhancement, spatial positioning and other fields.

The advantages of replacing human vision with machine vision are obvious.

(1) Training a qualified operator needs a lot of manpower and material resources, and the operator is prone to fatigue when working for a long time, resulting in low detection efficiency and accuracy. While machine vision only needs the initial hardware and algorithm investment, and high efficiency and high accuracy can be ensured even after long hours working.

(2) With the continuous improvement of industrial technology, it has become more and more difficult to detect the disqualified products from the appearance by human eyes, while machine vision can use a variety of detection means and high-precision sensors to achieve the rapid inspection of tiny features.

© Shanghai Scientific and Technical Publishers and Springer Nature
Singapore Pte Ltd. 2022
W. Mi and Y. Liu, *Smart Ports*, https://doi.org/10.1007/978-981-16-9889-7_7

(3)   The inspection with machine vision technology is indirect, without the need to set up a special manual detection point, which makes the inspection device largely free from space restrictions and capable of working in confined space or dangerous places that are difficult for human to reach. The configuration of inspection area is also relatively free.

7.1.1   Camera Vision Technology

The typical machine vision system includes two parts, image collection system and image processing system.

1.   Image collection system

The function of the image collection system is to convert the image and features of the measured object into the data processed by the computer. This system generally includes the light source, optical components and camera.

(1)   Light source

Light source and optical components are usually easy to be ignored, but the effect of these two parts on the whole image collection system cannot be ignored. The light source is used to influence the input of the visual system. For example, lighting a target can enhance the contrast between the target and the background, and highlight feature areas. Some features that are difficult to detect under normal light can be easily recognized in this situation. The subtractive color mixing is applied in the selection of light source, the feature area can be effectively highlighted by adding a filter and selecting a light source of different colors.

(2)   Optical components

Optical components generally refer to the lens, which can do some preliminary processing of incident light. For instance, in some fields of machine vision that need to calculate target parameters, the focal length of the lens should be optimized according to the object distance, the size of the image and the size of the sensor in the system, while in some applications of light wave collection involving specific wavelengths, corresponding filters need to be added to the lens.

(3)   Camera

Camera is a device that converts optical images into electrical signals. They are divided into CCD and CMOS cameras according to chip type; GigE, USB and Camera Link cameras according to interface type; color, gray and invisible light cameras according to imaging spectrum, and single-scanline and area-array cameras according to imaging mode. The image collection system should select the corresponding camera according to different objects and environments.

2. **Image processing system**

Image processing system consists of hardware and image processing software.

The hardware architecture of machine vision system can be divided into pc-based and embedded-system-based. The pc-based image processing system is independent of the image collection system, usually using high-performance industrial computer to debug and implement, achieving the desired accuracy and speed, and more complex functions. However, the disadvantage is that the development cycle is long. The whole system structure is complex and not suitable for copying, and it is suitable for the development platform of image processing software.

The embedded image processing system is usually integrated with the image collection system. The commonly used are based on DSP, FPGA and ARM structure. The embedded system is characterized by high degree of integration, simple use and installation, and low cost. As a product, machine vision system developed on PC is often transformed into machine vision system based on embedded system.

7.1.2 Light Detection and Ranging Vision Technology

Light Detection and Ranging (LiDAR), is a remote sensing method which measures the difference in information between emitted laser pulses and received reflected ones for contactless ranging and multi-dimensional scanning to measure the ranges of the surrounding environment. LiDAR technology is a combination of traditional radar technology and laser technology, which plays an important role in aerospace, military and mapping fields. Due to its small size, it has become a hot technology in UAV and IOT fields in recent years.

LiDAR is an active remote sensing system, which is based on laser ranging technology. It calculates the distance to the target point through the difference in the phase, amplitude and frequency of the emitted laser and the reflected laser. Laser has the characteristics of high directivity, high brightness and high coherence, which makes it a carrier naturally suitable for accurate ranging and velocity measurement, and has a longer detection distance than ordinary light. Because of the high speed and range resolution of laser, LiDAR has been widely used in environmental mapping, velocity measurement of moving object, object inspection and other fields.

LiDAR system is generally composed of laser transmitter, laser receiver and computer. The commonly used LiDAR is based on triangulation measurement and time of flight (TOF) measurement.

(1) Triangulation measurement

The principle of laser triangulation measurement is shown in Fig. 7.1, where the corresponding parameters are following.

F: The focal length of the lens;
S: The distance between the laser and the center point of the camera, which can be understood as the plane of the fixed camera and the laser;

Fig. 7.1 Principle of laser triangulation measurement

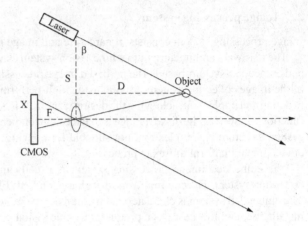

β: Laser angle;

X: The distance from the point at which the reflected laser passing through the camera focus intersects with the sensor CMOS to the point at which the parallel lines of the incident laser passing through the camera focus intersects with the sensor CMOS;

D: The distance from the object to the focus of the camera.

   To measure the distance D, first the laser ray is emitted to the object, then the reflected laser is projected onto the sensitive surface via the camera's pinhole imaging principle. At this point, the triangle formed by reflected laser, the parallel line of the emitted laser and the CMOS photosensitive plane is similar to the triangle formed by the emitted laser, the reflected laser and plane S. The following formula can be used to calculate the distance D.

$$D = F \cdot \frac{S}{X} \qquad (7.1)$$

   The laser triangulation measurement has high accuracy in the short distance. However, due to the pixel of the sensor, the farther the distance is, the lower the accuracy is. So, this method is only applicable to the short distance ranging within tens of meters.

(2)   TOF measurement

The TOF measurement is based on the already known speed of the signal, by calculating the signal transmission time between nodes for ranging. The measurement formula is below.

$$d = c * \Delta t \qquad (7.2)$$

where c is the speed of light, Δt is the difference of time measured in various ways.

The data source of TOF measurement is from the TOF of laser calculated based on electronic clock. Due to the influence of clock offset, the error is large in the condition of short range, so it is suitable for longer distance measurement.

7.1.3 Other Vision Technologies

In addition to the common camera vision technology and LiDAR technology mentioned above, other commonly used vision technologies also include structured light imaging technology, remote sensing technology, infrared thermal imaging technology, etc.

1. Structured light imaging technology

Structured light imaging technology is similar to the triangulation measurement in LiDAR technology, but the principles are different. The triangulation measurement relies on the pinhole imaging principle and calculates the distance according to the imaging points of the reflected laser on the sensor. While structured light imaging projects a pre-coded image onto the object and uses cameras set at different perspectives to pick up the reflection patterns. At this point, the pattern collected by the camera will be distorted due to the concave and convex surface of the object. By comparing the image collected by the camera with the original image, the 3D model of the object can be constructed. Compared with the laser triangulation measurement which can only scan the linear target, the structured light imaging can scan the opposite target at the same time without the need of additional rotating mechanism.

2. Remote sensing technology

Remote sensing technology is born and mature with the development of aerospace technology. This technology is to obtain ground information through optical, electromagnetic and radiation scanning with the help of space-based platforms such as aircraft and artificial satellites. In the 1970s, when remote sensing technology was born, it was mainly applied in the field of environment and resource exploration due to the limitation of computing power of the computers at that time. With the rapid development of computer technology in the 1990s, remote sensing technology has been rapidly applied in urban construction, military planning and other planning fields. At present, China has established a national environment macro-information service system, a disaster remote sensing monitoring and assessment operation system, a national spatial data infrastructure, and a 3D marine environment monitoring system based on remote sensing technology.

3. Infrared thermal imaging technology

Infrared thermal imaging technology is to convert thermal radiation that is invisible to the human eyes into images and patterns that can be observed by the human eyes by means of photoelectric detection. Compared with visible light imaging technology, infrared thermal imaging can work normally even under the conditions of rain, fog and lack of illumination at night, and has a strong environmental adaptability. In addition to the principle based on thermal imaging, this technology can also collect the temperature information of objects in the environment monitoring, and is widely applied in biological detection and machine inspection, non-contact temperature measurement, etc. At present, infrared thermal imaging technology has a number of applications in iron and steel industry, petroleum industry, electric power industry and other industrial fields. It is mainly used to monitor and prevent possible industrial accidents. In medicine, it can be used for the effective diagnosis of tumors, vascular diseases and skin diseases. In addition, infrared thermal imaging technology also plays a vital role in building inspection, forest fire monitoring, grain fire detection, combustible gas leakage detection and other fields.

7.2 Development of Machine Vision

7.2.1 General Introduction to Development of Machine Vision Technology

Machine vision technology is an interdisciplinary subject involving AI, computer science, biology, physics and other fields. In recent years, with the rapid development of AI science, machine vision technology, supported by intelligent computing hardware and ever-changing algorithms, has achieved huge economic benefits and social value.

In the 1970s, JVC introduced the first home camera with only 250 lines of visual capture. In 2004, Sony released the first HDV1080 HD camera with a resolution of 1440×1080. Nowadays, machine vision collection instrument has fully realized the low-cost 1080 P level, and cameras have gradually realized the collection of 2 K, 4 K and even billion-pixel levels.

With the cost reduction of visual collection instrument and the improvement of its performance, machine vision technology has developed rapidly. The purposes of visual instrument have gradually developed from the beginning of the simple replacement of human eyes to replacing the role of human mind, which is mainly due to the development of machine vision processing algorithms. Before 2012, machine vision algorithms were mainly designed by artificial feature engineering according to the actual scene and then entered into machine learning/mode identification classifier for training and detection. The machine vision algorithm after image dimensionality reduction required less computational power, but it had big limitations in the robustness of the scene and weak generalization ability. In 2020, Alexey Bochkovskiy took over from the original author Joseph Redmon and proposed YOLOV4, which greatly improved the accuracy and operating efficiency of the algorithm. In this context, the computational hardware of machine vision played the role of processing visual images and provided strong support for algorithms. With the development of high pixel of vision collection and the change of algorithm, the amount of data needed for machine vision image processing task becomes extremely large, which also puts forward higher requirements for machine vision computing hardware.

Since 2012, deep learning algorithms have gradually become the mainstream framework of machine vision. Traditional computing methods was unable to meet the requirement in computational data volume of deep learning, and GPU hardware began to promote the technological development of algorithms. From the beginning of dual GPU computing to the later Nvidia Titan X, they all rely on the GPU hardware conditions with large computing power. Among them, Nvidia dominates the current deep learning hardware market with its massively parallel GPU and dedicated GPU programming framework, CUDA, and its latest Quadro RTX 8000 delivers deep learning performance up to 130.5 TFLOPS. However, there are also more and more companies that have developed accelerated hardware for deep learning, such as TPU from Google, Knightslanding from Intel, GPU from AMD, GPU from Cambrian, etc. In addition, the high flexibility of FPGA also plays a certain role in deep learning computing, but it is not suitable for popularization due to its high difficulty in large-scale development, low overall performance cost ratio and low efficiency. On the other hand, in terms of hardware manufacturing process, the most mature and advanced architectures of 7 nm have been mass-produced, technical barriers in the manufacturing of 3 nm chip have been overcome, and the 2 nm chips are about to realize the mass production. Advanced hardware manufacturing technology will also greatly improve the computing power, leading to the rapid development of machine vision technology.

Visual collection instrument with high resolution is the natural foundation of high accuracy of the machine vision technology, and computer hardware configuration provides strong support for machine vision to realize visual data processing and analysis rapidly and accurately in effective time and within the data scope. The development and expansion of computer vision algorithm is the necessary condition for machine vision technology to be widely used in practice.

7.2.2 Machine Vision and Smart Traffic

In the twenty-first century, machine vision technology has developed rapidly, and has been widely applied and promoted in various fields. It has obtained huge economic benefits and social values. Common machine vision technology has covered agriculture, industry, medical care, transportation and other social fields. In agriculture, machine vision has been involved in every link from automatic picking of agricultural products to product quality inspection, which has laid a foundation for fine agriculture and automatic agricultural production. In industry, machine vision leads the development of industry from automation to intelligent manufacturing of Industry 4.0. Intelligent manufacturing combined with mechanical vision technology solves many problems in traditional manufacturing and plays an important role in improving social production efficiency and reducing labor intensity of workers. With the continuous development and maturity of machine vision technology, its impact on the production and manufacturing industry will be more enormous. In medical care, machine vision has a mature application in drug production and medical auxiliary diagnosis. The processing technology of medical image has reached the expert level, which has far-reaching significance for promoting the primary medical care.

In traffic, according to data published by United Nations that by 2050, nearly two-thirds of the world's population will be living in big cities. The wave of urbanization will put great pressure on the city, especially in the aspect of traffic management. Machine vision technology can replace human eyes to judge and measure, and effectively relieve the pressure of traffic management. For example, "Haiyan", "Kunlun" and other systems in China can well realize the comprehensive control of road traffic, improve traffic conditions, and provide guarantee for the safe travelling of citizens. The application of machine vision in traffic management can be divided into traffic incident detection, vehicle safety assurance, vehicle license plate recognition and autonomous driving technology.

(1)   Traffic incident detection

Traffic incident detection mainly uses machine vision technology to realize the image analysis of traffic sequence, target identification and tracking for pedestrians and vehicles. On this basis, the traffic scenario is understood, so as to capture violations such as running a red light, converse running, illegally changing lanes, making phone call while driving, not wearing a seat belt and so on, and realize automated monitoring, video recording and alarm of traffic incidents and traffic congestions. With the help of machine vision technology, traffic management departments can realize intelligent traffic incident detection and off-site law enforcement.

(2)   Vehicle safety assurance

Machine vision can assure the vehicle safety by means of visual enhancement and extension in the process of driving. Visual enhancement is to monitor the external environment with visual collection instrument, and enhance the effect by certain technical means to achieve real-time monitoring of the surrounding traffic environment. In addition, machine vision technology can also enhance the visual environment under different climate and time conditions, improve its visual effect, and achieve driving assistance under adverse conditions such as low illumination and low visibility. Visual extension can break the light limitation of human eyes and extend sight in the blind area in the process of driving by visual sensor, and make visual compensation for the driver. One of the most common visual extension safety means is the reversing image system, including the early radar protection to the present 360° panoramic protection, which has undoubtedly greatly extended the visual field of the driver and ensured the safety of the vehicle.

(3)   Vehicle license plate recognition

Vehicle license plate recognition (VLPR) technology is a hot topic in modern traffic system research. The license plate no. of the vehicle is fixed, so all the information of the corresponding vehicle can be found by using the machine vision technology to recognize the license plate no. VLPR is an important step to achieve intelligent traffic management, and has been widely used in vehicle management in the airport, port, residential area, and non-stop electronic toll collection and other fields.

The early traditional VLPR mainly included image collection, image preprocessing, license plate positioning, character segmentation and character recognition. Nowadays, with the development of AI technology, end-to-end VLPR based on deep learning has been gradually promoted.

(4) Autonomous driving technology

Machine vision technology has also been incorporated into driving technology, known as autonomous driving. The application of machine vision in autonomous driving mainly has two aspects, namely obstacle detection, road mark detection and license plate identification.

In autonomous driving, the appearance of objects cannot be known in advance, so the timely identification of objects is of great significance for safe driving. At present, there are three main object detection algorithms based on machine vision: feature-based, optical-flow-field-based and stereoscopic-vision-based. The object detection technology based on stereoscopic vision does not need to know the shape, size and other information of the object, and does not need to consider whether the object moves, but also can get the specific position of the object, which has become the mainstream direction for research.

In autonomous driving, road boundary detection is related to whether the vehicle can drive according to the traffic rules. Many countries have developed their own road recognition and tracking systems based on vision, including LOIS system, GOLD system and RALPH system. Road boundary detection and lane recognition include two methods, one is feature-based recognition method, and the other is model-based recognition method. The feature-based recognition method detects the road boundary from the captured image according to the actual road situation and the actual road features (color features, gray features, etc.).

In addition to the most common applications of machine vision in the traffic field as mentioned above, Nvidia in 2019 launched a "mobile city" data set, which covered the camera data of a city, aiming to analyze the most advanced machine vision algorithms and spatiotemporal analysis algorithms, to complete the intelligent management of the whole urban traffic flow.

In the field of intelligent traffic management, machine vision technology plays a vital role in alleviating traffic pressure, maintaining traffic order and improving traffic safety. In the future, with the development of machine vision technology, intelligent traffic management is developing towards the direction of intelligent management without manual intervention.

7.3   Applications of Machine Vision Technology in Smart Port

7.3.1   Early Applications of Machine Vision Technology in Port

1.   Application of camera vision technology in port

Machine vision was originally applied to solve the problem of automatic identification of objects in labor-intensive industries. In port operations, each container handling operation, import collection and export delivery need to inspect various parameters of the container body. New intelligent port facilities have been able to solve these problems in the way of machine vision.

For example, in the anti-hoisting system of chassis, it is traditionally necessary to deploy personnel in the process of container handling to check on site the twistlocks between the container and the chassis frame, the bottom locks of the container, etc., to prevent false hoisting. Now, machine vision detection has replaced the manual inspection. In 2019, the anti-hoisting system developed by Qingdao automated terminal in China has, for the first time in the world, realized automatic container collection operation at the land side, which has avoided manual intervention in the process of container collection, improved operation efficiency and further enhanced safety.

2.   Application of LiDAR technology in port

Since Maiman developed the first ruby laser in the 1960s, a large number of LiDAR technologies have been used in various aspects of the field of environment perception. From the 1D laser rangefinder first used in the military field in the 1960s to the 2D LiDAR widely used in environmental exploration since the 1990s, with the development of the IOT technology in recent years, small LiDAR instruments that are capable of 3D scanning have gradually been widely used.

In recent years, with the development and construction of "One Belt and One Road" in China, and the development and application of the new technologies, the port industry is undergoing the transformation of automation and intelligence. Due to the large amount of transportation involved in port operations, the introduction of a large number of automation technologies can effectively reduce the errors caused by human fatigue, and through the centralized planning of automated transportation equipment, the operation efficiency can be significantly improved, among which the automated guided vehicle (AGV) technology is the key of intelligent transportation of port. AGV is a kind of battery-driven container transportation vehicle that uses a series of automatic technologies such as automatic navigation, unmanned driving and automatic obstacle avoidance. The unique positioning and route planning system is the core technology of such kind of vehicles.

For early AGVs, ground magnets were used as navigational reference, and electromagnetic signals were used to determine the spatial location of the AGV and plan its route. The new generation of AGVs began to use the LiDAR scanning guid-ance method based on the construction of 2D LiDAR, which uses on-board LiDAR to continuously scan the surroundings to create a local map while the AGV is in motion. With Kalman filtering algorithm and other technologies, the local map and the overall map reserved in the AGV system can be used to obtain the actual motion nodes of the AGV, which can be used for real-time correction of the AGV route. Compared with the magnet method, it has a higher precision. Meanwhile, due to the real-time charac-teristics of LiDAR measurement, it can also be used as an anti-collision system when the AGV is in motion. As the LiDAR establishes a local map, it can scan the obstacles on the route and stop to alarm or choose to bypass. Laser ranging technology is also used in the field of positioning. Quay crane is the main quayside operation facility in container terminal, which is used for loading and unloading containers in-between ship and quay. The operations of quay crane include lifting container from chassis or landing container onto chassis, which traditionally relies on the driver to complete the alignment of the chassis and the spreader by visual inspection. The operation is difficult and inefficient, and collision between spreader and chassis is easy to cause damage, which brings many safety risks.

In 2016, Qingdao Qianwan Terminal in China developed an automatic positioning system for chassis under quay cranes based on the principle of laser scan ranging, which applied a laser scanning system to range for the chassis under quay crane. With the microprocessor, the shape of the container, height and its offset distance to the lifting point of the quay crane are identified, and the prompt message is displayed in LED screen to remind the driver to adjust the parking position of the chassis. After testing, the introduction of LiDAR technology has increased the overall efficiency of the operation by 9.74%, and greatly reduced the safety risks caused by the collision of the chassis and the spreader.

3. Applications of other machine vision technologies in port

Other machine vision technologies are mainly structured light imaging technology, remote sensing technology, infrared thermal imaging technology and so on. Such technologies have a narrow application field, and their applications in port are mainly in the following aspects.

(1) Structured light imaging technology

Structured light imaging technology also has the function of space position calcu-lation and object surface information detection as LiDAR technology. Similar to LiDAR, structural light imaging can also be used in port for automatic lifting adjust-ment of spreader and container, scanning of container surface damage and so on. In addition, the container position and orientation measurement method based on the structured light decoding technology uses the structured light transmitter and optical sensor installed on the spreader to calculate he position and orientation relationship of the container relative to the spreader via characteristic coordinates. Compared with the LiDAR system, this system provides multi-dimensional scanning capability without complex rotating mechanism, and the response time is shorter. It can realize automatic orientation control during the lifting process.

(2)   Remote sensing technology

Remote sensing technology is an important means of channel detection in modern ports. Due to the hydrological movement in coastal and estuarine areas, the shoreline and underwater environment are often in the process of constant changes. Traditionally, large-scale hydrological surveys have to be organized regularly to update the hydrological and sediment data in the records in view of the changes in the waterway, which is time-consuming and laborious. With the help of satellite remote sensing technology, the real-time information of waterway can be acquired conveniently, quickly and at low cost. Information such as water depth and water quality can also be obtained through the analysis of visible spectrum collected by satellite remote sensing technology.

With the development of space technology and computer technology, the resolution of remote sensing technology has been effectively improved. Nowadays, remote sensing technology has been applied in a large number of engineering fields, including its application in port planning and management. High-resolution remote sensing technology can identify and classify buildings, blocks, roads and large facilities in port, as well as identify ships in port, providing a reference for resource planning among ports.

(3)   Infrared thermal imaging technology

In port monitoring system, infrared thermal imaging technology is often used in warehouse anti-theft and security system due to its all-time characteristics. The introduction of this technology enables the monitoring system to monitor and record the flow of people in the port area normally at night, which has greatly reduced the human resources allocated to guard the warehouse area, and been connected to the alarm system to provide certain evidence for the security.

Infrared thermal imaging technology can also be applied in ship searching at night in port. In 2009, the Ravenna Port in Italian purchased a batch of multi-sensor thermal imaging cameras for the coast guard and port pilot institutions. With the help of these instruments, the coast guard's monitoring range has been extended from the port front to the adjacent coastal ports, and nearby ships can be identified and searched normally, no matter in days or nights, ensuring the safety of the surrounding environment of the port.

7.3.2   Typical Applications of Machine Vision in Smart Port

The development of automated terminal is the key to promote steady and rapid development of global economy. Nowadays, automation technology is developing rapidly in the world, and people have a deeper understanding of machine vision technology. The application of machine vision technology has made it possible to work in dangerous environment in port, which is not suitable for manual operation, and greatly improved the operation efficiency and product precision. With the maturity and development of machine vision technology, its application scope is more extensive. The typical applications of machine vision technology in the construction of smart port can be divided into three categories, mainly including image recognition, target positioning and security protection, which can basically summarize the application condition of machine vision technology in port intelligence.

7.3.2.1 Application of Image Recognition in Smart Port

As the most traditional visual detection technology, image recognition is to recognize targets and objects of various modes by image processing and analysis mainly based on the machine vision technology. In the construction of smart port, image recognition technology has played an important role mainly in container no. recognition, container information recognition, chassis no. recognition and so on.

1. Container no. recognition

As the first machine vision system applied in the terminal, the container no. recognition system mainly uses OCR optical identification technology in the traditional field to recognize and collect the container no. At present, the container no. recognition system has been widely promoted in Tianjin Port, Taicang Port, Xiamen Port, Lianyungang Port and other ports in China, and become one of the indispensable machine vision systems in the construction of smart port.

As shown in Fig. 7.2, the container no. follows the international standard ISO 6346, which is composed of three parts: 4 capital English letters for container owner,

Fig. 7.2 Container no.

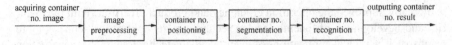

Fig. 7.3 Container no. recognition process

6 digits for container classification and 1 check code. These 11 ISO characters are the unique identification code of the container, which plays a very important role in the intelligent management of container. Compared with the previous handling operation process of quay crane, the collection of container no. mainly relies on manual identification and manual input. Facing such an odious environment of quay crane operation, it is not only inefficient but also with safety risks.

The process of container no. recognition system mainly includes container image collection, image preprocessing, container no. positioning, container no. segmentation and container no. recognition, as shown in Fig. 7.3. The outdoor specialized HD IP Camera is generally adopted in container image collection, and it is installed on the saddle beam and the legs of the RTG, in order to obtain clear image for subsequent processing.

Image preprocessing is a key step for the subsequent recognition of container no. In the process of container image collection, it is often affected by weather, uneven illumination and collection direction, and the main problems include noise pollution, blurry image, tilted container, interference of character position and so on. In order to improve the speed of image processing and make the operation more convenient, the colored image is grayed first. For noise pollution, smoothing filter can be effective for image denoising, and the commonly used include the nonlinear and linear smoothing filter. Linear filters include box filter, mean filter, Gaussian filter and so on, while nonlinear filters include median filter and bilateral filter, both of which have good effect on suppressing the image noise. In order to highlight the "useful" information in the image, the differences between different target features in the image are enhanced, to improve the precision of target extraction. Commonly used image enhancement methods can be divided into two kinds, one is direct contrast enhancement method, and the typical is gray level transformation; the other is indirect contrast enhancement method, and the commonly used are histogram stretching and histogram equalization.

Container no. positioning is to use a series of image processing methods to remove the non-target area and extract the target area of interest. Character positioning methods of container no. can be divided into three categories. The first is based on character texture features, the second is based on character structure features, and the third is based on character edge features. Although the characters of the container no. are extracted accurately, the characters are mostly hand-printed, and the way of writing and the shape of the container body determine that the container no. will have a certain tilt, and the image of the tilted container no. will affect the recognition result. Therefore, tilt correction is a very necessary step, and common tilt correction methods are usually PCA-based, project-based, Hough-transform-based, nearest-neighbor-based and so on. Each method has a good performance in terms of correction accuracy and real-time performance, but there is a big gap between their robustness and the requirements of practical application in the face of different actual tilt conditions.

After the accurate positioning of the container no., it is necessary to segment the characters of the container no. to accurately recognize it. The quality of the segmentation results directly affects the result of the character recognition. At present, the methods of container no. character segmentation include horizontal and vertical segmentation, character segmentation based on nonlinear clustering, segment confidence-based binary segmentation, separator's symbols frame of reference, etc. The single character after segmentation has to be recognized and there are two main recognition methods. One is character recognition based on traditional methods, such as template matching, feature matching, feature analysis matching, etc., and the other is character recognition based on neural network, such as BP neural network, artificial neural network, deep learning, etc. Finally, the recognition result of the container no. is sent to the terminal management system.

The application of container no. recognition system has promoted the development of the automation and intellectualization of ports. Combined with the TOS, ECS systems, it can realize intelligent recognition and intelligent tally in the process of loading and unloading operation in the fully automated terminal, and has changed the traditional mode of manual confirmation, improved loading and unloading efficiency and security of the smart port, reduced the operating costs, and become one of the most essential systems in automation and intellectualization of operations in smart port.

2. Container information recognition

In the process of automation, automated visual inspection technology has been gradually introduced to container terminals, using image processing technology to rapidly recognize container information, including key information such as the door orientation, seal, dangerous goods and so on, which has improved the efficiency of the ports while reducing manpower input and the labor intensity. The container information recognition system has been put into use in Ningbo Zhoushan Port, Tianjin Port, Xiamen Port, Taicang Port and other ports in China.

(1) Container seal recognition

As shown in Fig. 7.4, the lead seal of a container is the safety lock of a container, the intact seal represents that the container has not been opened. The seal commonly used in the shipping process of containers is the traditional lead seal of high protection,

Fig. 7.4 Container lead seals

which is composed of a lock body and a lock bolt. The lock body is the information carrier, generally printed with a number of codes or bar codes, and commonly cylindrical. When the lead seal is broken, it means the container has been opened.

In the process of inspecting lead seals by using image recognition, it is very difficult for the traditional lead seals because of the characteristics of small volume, various colors, various forms and installation positions. Usually, the recognition rate of machine vision for the traditional lead seal is only about 60–70%.

Since traditional lead seal is not conducive to the development of port automation and intellectualization, and in order to improve the fault tolerance performance and recognition rate to lead seals, a new type of QR code lead seal arises at the historic moment. It inherits the characteristics of traditional lead seal and has the function of encryption processing. Meanwhile, the QR code seal is easier to be captured and recognized by the vision camera, and it can also be recognized when the code is partially missing, which has better fault tolerance than the traditional seal.

(2)   Container door orientation recognition

Container door orientation recognition is an additional work of automated port. In the process of container collection transportation, if the container door is facing the chassis cabin, it will bring inconvenience to the cargo inspection and container handling in the port. Therefore, at present, large automated terminals in China have made requirements on determination of container door orientation, and also arranged special equipment to adjust the door orientation.

In the automated terminals, door orientation detection is usually accompanied by OCR recognition system, and the door orientation detection is carried out at the same time as the container no. recognition. The principle of door orientation detection is mainly to analyze the different imaging characteristics of the container doors and the container rear. As shown in Fig. 7.5 are the actual photos of the container doors and container rear.

Fig. 7.5 Container doors and container rear

Fig. 7.6 Dangerous goods label on the container

(3)   Dangerous goods label recognition

The dangerous goods label of container shows the attributes and characteristics of the goods packed in the container. However, due to the multiple circulation of containers with dangerous goods through long-distance transportation, handling, transition and so on, it is very easy to cause the dangerous goods label of the container to fade, damage and fall off, and even some packing companies may forget to post the dangerous goods label, which will bring some safety risks to the operation. In order to reduce the hidden danger, it is necessary to recognize the dangerous goods label. Figure 7.6 shows the dangerous goods label on the container.

3.   Chassis no. recognition

Chassis are important means of transportation in the container terminal connecting the yards and the quay. The tracking of the chassis depends entirely on the observation of the driver and the personnel on site in the terminal. Although this method can achieve the tracking of the chassis information, it has disadvantages such as low efficiency and manual errors. Therefore, in order to improve the handling efficiency of container terminal, it is necessary to recognize the chassis no.

At present, the chassis no. recognition is mainly achieved with vision camera. According to the real-time collection of video streaming images, the real-time recognition of the chassis no. is realized with OCR technology. The recognition performance of the chassis no. is shown in Fig. 7.7.

7.3.2.2   Application of Target Positioning in Smart Port

Image processing and computer vision in port automation research, are always problems worthy of studying. The accurate positioning of the target plays a very important role in target recognition as well as image analysis and processing. The typical applications of target positioning in the port automation include chassis positioning, train positioning, container positioning, etc.

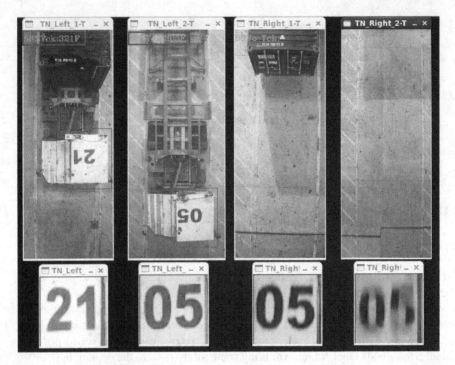

Fig. 7.7 Recognition performance of the chassis no.

1.   Chassis positioning

The facility connecting the container terminal yard and the front quay crane is the chassis. Traditionally, the tracking and positioning of the chassis depend on the experience of the driver and the personnel by the quay crane. In order to solve the problems of manual error and low efficiency, and improve the automation level of container terminal, machine vision technology is often used to position and track the chassis. At present, the vision-based chassis positioning system has been used in many ports in China, including Qingdao Port, Xiamen Port, Shanghai Port, Tianjin Port, Guangzhou Port and other ports.

At present, some ports have also tried other positioning technologies, including the photoelectric sensor based, or sticking the reflective plate on the chassis beam to detect the chassis position through the infrared beam reflection. The GPS based technology is to realize accurate positioning of the chassis by GPS locator installed in each chassis. All the above positioning methods can achieve certain positioning accuracy; however, they cannot effectively solve the problem of accurate positioning. At present, the chassis positioning systems based on LiDAR and machine vision technologies are relatively in wide use.

The main process of the chassis positioning system based on machine vision technology includes image collection and preprocessing, moving target detection, feature extraction, feature matching and data transmission. When the chassis enters the operating bay under the crane, the industrial computer recognizes the chassis no. with image processing technology. The size information of the container is obtained by combining the bay no. and the chassis no., and then the preset position of the target feature under the crane is determined. The outdoor high-definition camera is triggered to obtain the scene image and then the image processing software is used for analysis. The target is inspected to obtain the real-time position of the target feature and transmitted to the host computer. The host computer matches the real-time position of the target feature and the theoretical position to obtain the deviation direction of the target feature and the theoretical position, and transmits the analysis results to the display board. If the error exceeds the prescribed range, the driver will be informed through the display board to adjust the parking position until the error reaches the prescribed range. If the error reaches the prescribed range, the chassis positioning will be shown to be successful. Figure 7.8 shows the actual performance of chassis visual positioning.

The LiDAR-based chassis positioning system uses laser scanner to scan the container on the driveway, to determine the driving direction of the chassis and the location of the cab. The background processor control system can locate the chassis according to the central position of the container on the chassis, display the chassis positioning information and provide it to the driver of the chassis, so that the container and the spreader as well as the container and the unloaded chassis can be aligned in advance. Figure 7.9 shows the cloud map of the scanning points of the LiDAR-based chassis positioning system.

2.   Container positioning

Containers are the main handling objects in container terminals. In traditional terminals, the handling operations are achieved by the manual operation of the crane driver to grab the containers. With the advancement of automation development process,

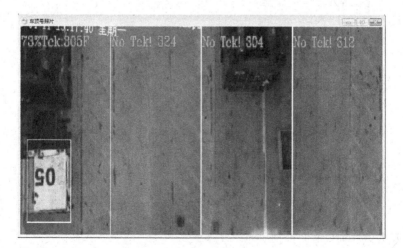

Fig. 7.8 Actual performance of chassis visual positioning

Fig. 7.9   Cloud map of the scanning points of the LiDAR-based chassis positioning system

such large automatic handling facilities as automatic quay cranes and automatic RTGs appear in succession. The detection of container position by machine vision, which is the substitute of eyes, is the prerequisite of automatic handling operations. In the entire automatic handling process, the trolley system does not require human participation, so the actual location of the target container must be obtained. The container positioning system is shown in Fig. 7.10.

This technology is mainly realized by LiDAR scan ranging. At the beginning of the automatic operation, the LiDAR scans the 3D point cloud data of the storage yard

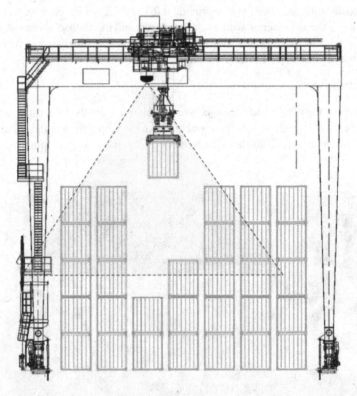

Fig. 7.10   Container positioning system

to detect the height of the front container row and the minimum distance between the adjacent container and the current container row, so as to avoid collision.

7.3.2.3   Applications in Safety Protection in Smart Port

Safety protection is the premise of port security. At present, with the rapid growth of port throughput, the security pressure of port operation is increasing. In the process of continuous operation, economic and personnel losses caused by unsafe behaviors of human, unsafe condition of things and bad environment occur from time to time. The typical applications in safety protection in port automation include anti-hoisting detection, anti-collision detection, detection of foreign object on tracks, human detection, etc.

1.   Anti-hoisting detection

No matter in the traditional terminals or the intelligent terminals, there are security risks that the locked chassis is hoisted during the operation. In order to reduce the occurrence of such accidents, in most traditional terminals, it relies on manual on-site inspection, judgments are made according to manual observation and the experience of operating drivers, which can greatly reduce the occurrence of such accidents, but greatly increase the labor cost. At present, the container lifting pictures from the remote monitoring cameras are connected to the driver's cab, so that the driver can constantly observe the real-time picture to understand the lifting situation of the chassis, so as to avoid the occurrence of dangers. This way, on the contrary, increases the working intensity and complexity of the driver, and the security risks increase. In the automated terminal, LiDAR is used to continuously scan the container and chassis to determine whether they are separated, so as to realize the anti-hoisting function of the chassis. However, LiDAR alone cannot deal with all kinds of chassis hoisting. In order to better meet the requirements of safety protection in port operation, the anti-hoisting system of chassis combining machine vision and LiDAR is adopted. Whether the container is completely separated from the external chassis is determined by intelligent recognition through the lock recognition of the chassis, real-time tracking of the chassis frame and laser scanning. If it is detected that the container is not completely separated from the chassis when hoisted to a certain height, an alarm and stop signal will be output, to realize intelligent, unmanned and dynamic monitoring and management. At present, the system has been applied in Ningbo Zhoushan Port, Shanghai Port, Tianjin Port, Guangzhou Port, Dalian Port and other ports in China.

The anti-hoisting system based on machine vision technology is mainly composed of hardware and software. The hardware mainly includes detection camera and LiDAR, industrial computer, display screen, alarm and confirmation button in the main control room. Each of the four detection cameras is responsible for photographing the two locks at one end of the chassis, and the LiDAR is used for 3D scanning. The image processing module is responsible for the recognition of image data and the transmission of processed results. The LiDAR data processing module is responsible for processing LiDAR data and sending results. The industrial control and

Fig. 7.11   The wheel tracking alarm performance of anti-hoisting of chassis

display machine is responsible for communication with PLC and image processing module, and is placed in the cab for the driver to confirm at any time. The confirmation button is linked to PLC and is used when the driver needs to confirm manually. The main process of the software includes confirming chassis location, collecting the benchmark image, extracting the feature points of chassis frame, collecting new image, extracting the feature points of chassis frame, matching the feature points and outputting the results.

Feature point matching algorithm is adopted to extract the feature points of the chassis frame. At present, the feature point detection algorithms commonly used in the field of machine vision mainly include SIFT, SURF and ORB, among which the ORB algorithm combines FAST corner detection with the local binary feature descriptor BRIEF, and its real-time performance is superior to SIFT and SURF. After feature extraction, feature matching is needed to detect whether the two have commonness. The commonly used feature matching methods include K-nearest neighbor (KNN) matching, Brute-force matching, FLANN matching, etc. After feature matching, within a certain error range, safety condition can be assured; otherwise, it is mistakenly hoisted. The wheel tracking alarm performance of anti-hoisting of chassis is shown in Fig. 7.11.

The anti-hoisting system of chassis based on LiDAR, uses LiDAR technology to achieve anti-hoisting function. The system processes the data points scanned by LiDAR in real time, identifies the container side, and detects the container landing signal as well as locking and unlocking signal of the crane. Meanwhile, with the continuous operating of the hoisting mechanism, the system will automatically calculate the hoisting altitude change, and compare it with the threshold value. When it is greater than the threshold value, the system will start to work. Once the anti-hoisting algorithm works, the system will detect whether the container and the chassis are properly separated. Figure 7.12 shows the point cloud data of anti-hoisting system of chassis based on LiDAR.

Fig. 7.12   Point cloud data of anti-hoisting system of chassis based on LiDAR

2.   Anti-collision detection

As a logistics distribution center, the main operating facilities in the port include RTGs, stackers, chassis handling equipment, forklifts, etc., which cannot operate independently and always need the participation of on-site personnel and the mutual cooperation between facilities. Often due to the influence of light and visual field factors of the drivers of these large facilities, the collision of facilities and personnel on the site is caused, resulting in serious accidents.

In order to prevent the occurrence of the above accidents, machine vision technology is also widely used in the anti-collision detection of gantries. At present, the detection methods of anti-collision detection of gantries based on machine vision mainly involve LiDAR and camera vision, whose principle is to use point cloud data of LiDAR or 2D image data to detect, recognize and early warn foreign bodies on the route of the gantry.

(1)   Anti-collision based on LiDAR

LiDAR is usually installed at the four legs or beams of operating facility. The point cloud data in the protected area are obtained through laser scanning, and then stitched and restored to a 3D model or 2D image, so as to further analyze the image features and obtain important data such as the size, contour and distance of foreign bodies in the protected area.

As an active vision, the advantage of LiDAR lies in that it is less affected by the external environment. No matter what the external lighting conditions are during the day and night, it can maintain stable operating ability.

(2)   Anti-collision based on vision

Anti-collision based on vision can be divided into two categories: monocular anti-collision and binocular anti-collision. Commonly used monocular vision often realizes the detection and recognition of the obstacles in the given area through the feature analysis of the image itself, and realizes a pseudo-3D distance measurement by setting the calibration area. Binocular vision is the same as LiDAR to obtain the actual distance between the obstacle and the facility, as well as the 3D size and depth information of the obstacle.

Anti-collision based on vision is greatly affected by illumination, so it is often necessary to add supplementary light equipment to ensure the illumination stability

Fig. 7.13 Configuration of anti-collision area of gantry

of the external environment. Compared with LiDAR, anti-collision based on vision has clearer images of obstacles, and the types and shapes of obstacles can be distinguished. Meanwhile, vision camera also has the advantage of low cost compared with LiDAR.

As shown in Fig. 7.13, the results of anti-collision of gantries based on the two modes mentioned above are mainly divided into deceleration area and stop area. When the machine vision recognizes that the obstacle is located in the deceleration area of the traveling route, the gantry will give an audible alarm and send deceleration instructions to the control system of the gantry to decelerate. When the machine vision recognize that the obstacle appears in the stop area, it indicates that the obstacle is very close and the situation is very dangerous. The identification system will immediately send the stop instruction, and control the gantry to brake and stop driving.

3.   Detection of foreign bodies on the tracks

At present, the rail-mounted cranes have become the widely used handling equipment in terminals of the world, such as RTGs, QCs and so on. As the name implies, the rail-mounted cranes are mainly driven in accordance with the established tracks. Once there are hard foreign bodies on the tracks, such as stones, iron, etc., it is easy to cause derailment and other dangerous situations. Meanwhile, foreign bodies will also cause great damage to the tracks and equipment. Therefore, detection of foreign bodies on the tracks is also a necessary means of safety measures in terminals. The traditional track detection mainly relies on the staff to inspect the tracks regularly. The foreign bodies can be found with this method, but not in a timely manner, and there are still some safety risks. On the other hand, it is a waste of manpower, material and financial resources. As shown in Fig. 7.14, the tracks of a QC in a Chinese terminal are cleaned manually on a regular basis.

In recent years, with the rapid development of machine vision technology, the detection of foreign bodies on the tracks also begins to rise rapidly with the help

Fig. 7.14 The foreign
bodies of the QC tracks are
cleaned manually on a
regular basis

of machine vision technology. At present, detection of foreign bodies on the tracks is mainly realized by the traditional visual detection technology, that is, using the camera to shoot the track picture, and using the method of Hough transformation for detecting line to extract foreign body images; or direct detection of foreign bodies with machine learning method of pattern recognition; or using the frame difference method to extract and update the image background, and calculate the height and length of the foreign body by continuously shooting two frames of the same position of the track.

Under the protection with machine vision technology, the security system can realize real-time and accurate detection of foreign bodies on the tracks round the clock, automatic alarm and early warning to prevent accidents before they happen.

4. Humanoid detection

There are a large number of heavy machineries in the operation area of the terminal. In order to prevent casualties caused by unauthorized entry of personnel during operation, most terminals adopt traditional video monitoring method, with which cameras are installed in the areas that need monitoring, and professional personnel are arranged to observe the monitoring images of the cameras. However, this approach also has many disadvantages, such as the failure to prevent the occurrence of dangerous behaviors in real time, the missed and false alarms of the monitoring system, the difficulty of data analysis and the long response time. The method of machine vision is adopted to realize intelligent video surveillance. Without human

Fig. 7.15 Illustration of humanoid detection system in the port

intervention, the machine can analyze the video image source in real time and automatically, identify and obtain the key information, realize the positioning, recognizing and tracking of the target in the dynamic scene, and make the corresponding response at the same time. The humanoid detection system is used in almost all ports.

Humanoid detection system consists of video acquisition unit, humanoid detection unit, alarm output unit, etc. Video acquisition unit is responsible for acquiring images from the camera or other system, and providing them to humanoid detection unit, which is the core of the system, extracting the humanoid feature information from the image provided by the video acquisition unit. At present, methods such as HOG and Kalman filter, Hu moment and Zernike moment, wavelet transform and dynamic frame are used to extract the humanoid feature information in the image. After the humanoid feature extraction, SVM is used to recognize and classify the humanoid, and the detection results are provided to the alarm output unit. According to the detection results provided by the humanoid detection unit, the alarm output unit will determine whether the person in the scene image is included in the monitoring list. If so, it will perform the preset alarm steps. As shown in Fig. 7.15, the protection is provided by the humanoid detection system in the port.

With the development of machine vision technology, it has gradually involved in the construction of automated terminals from intelligent manufacturing and intelligent transportation. No matter the mainstream camera vision, radar vision, infrared thermal imaging, remote sensing imaging, structured light and other new or old technologies, have found their roots in the terminal construction and developed rapidly. At present, the applications of machine vision in smart ports in Chinese market are mainly in image recognition, target positioning and safe protection, covering container, container trucks, automated handling equipment. The machine vison technology is involved in the whole process of handling operations in the port, to effectively monitor, detect, secure and guide the handling operations of the smart ports.

Bibliography

1. Szeliski R (2010) Computer vision: algorithms and applications. Springer Science & Business Media
2. Mi C, He X, Liu H et al (2014) Research on a fast human-detection algorithm for unmanned surveillance area in bulk ports. Math Probl Eng 2014
3. Mi C, Shen Y, Mi W et al (2015) Ship identification algorithm based on 3D point cloud for automated ship loaders. J Coast Res 73(10073):28–34
4. Mi C, Zhang Z, He X et al (2015) Two-stage classification approach for human detection in camera video in bulk ports. Pol Marit Res 22(S1)
5. Szegedy C, Vanhoucke V, Ioffe S et al (2016) Rethinking the inception architecture for computer vision. In: Proceedings of the IEEE conference on computer vision and pattern recognition, pp 2818–2826
6. Mi C, Zhang Z, Huang Y et al (2016) A fast automated vision system for container corner casting recognition. J Mar Sci Technol 24(1):8

Chapter 8
Smart Port and Virtual Reality/Augmented Reality Technology

8.1 Introduction to Virtual Reality/Augmented Reality

Virtual reality (VR) and augmented reality (AR), as the hottest technology nowadays, have been closely related since their birth. VR means that all the content of the whole scene is virtual, and not associated with reality, just as we enter a 3D game, while AR means that there is the presence of a large number of realistic contents in the vision field, on the basis of which virtual contents are superimposed. The two parts can realize interaction, just like a cup appears suddenly on a table, but it is not real. People are often confused by VR and AR, which has motivated the development of mixed reality (MR). MR is a virtual reality technology, which combines VR with AR to build an interactive feedback loop among the virtual world, the real world and the users, so as to enhance the sense of reality of user experience. Physical entities and digital objects can interact in the new visual environment.

There are three categories of VR devices, namely mobile phone VR box, VR headsets and PC connected VR. The mobile phone VR box is cheap and can be used to watch VR movies and play some simple games with the mobile phone, and the representative product is Google Cardboard. The product structure is very simple, even can be made manually, and the working principle is to use the vision difference of left and right eyes, so as to produce a 3D feeling. The VR headsets and PC connected VR are the most promising VR devices with a sense of technology, as shown in Fig. 8.1.

At present, the common VR headsets and PC connected VR mainly consist of the following four parts: head-mounted displayer (HMD), host system, tracking system and controller. HMD, commonly known as virtual glasses, is worn by the user to realize the optical function of AR/VR. Host system refers to all kinds of equipment to assure the normal operation of HMD, such as smart phones, PC, etc., whose performance will greatly affect the display effect of VR/AR. The tracking system, typically a peripheral equipment of the HMD, or integrated into the HMD, includes gyroscopes, sensors and magnetometers. Its core function is to capture the user's movements to rotate to create immersive experiences, such as when you look up,

W. Mi and Y. Liu, *Smart Ports*, https://doi.org/10.1007/978-981-16-9889-7_8

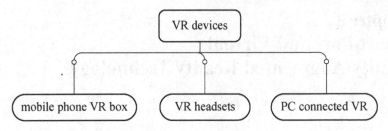

Fig. 8.1 Categories of VR devices

you can see the sky. The controller is typically a handheld device that tracks the user's gestures. The HMD includes the displayer, processor, sensor, camera, storage, battery, lens. Most displayers have one or two screens, and the one with 4 K resolution has become the dominant choice. The separate HMD devices use organic light emitting diode (OLED) screens, while the integrated HMD devices use microprojection technology. OLED has many advantages over LCD, such as a faster refresh rate and a lower latency, which ensure that users won't get dizzy during use. Microprojection is the most widely used in AR devices, such as in Google Glass and will be widely used in future AR devices.

The processor is the core of the device, which is used to generate images and calculate attitude positioning based on data from the gyroscope. In order to avoid dizziness, the image refresh rate should reach 90 Hz, which requires a lot on calculation speed, and only a good chip can do the job. Sensors track the movements of the user's eyes and head, and send information to a processor that produces an image. Only sensitive and efficient responses can create an immersive experience for users. Sensors include FOV depth sensor, gyroscope, accelerometer, magnetometer, etc. Cameras are used for some HMD devices to take pictures, track location, and map the environment. As the name implies, the storage system provides storage space for video and image of high revolution, which needs fast reading and storing. Batteries provide power to work for suitable hours. Most of the lenses are aspheric.

It's likely that AR is an upgrade of VR, but to achieve that upgrade, the two technologies have gone to completely different paths in terms of optical structure. The optical structure of VR (shown in Fig. 8.2) is relatively simple. In fact, it is a simple convex lens imaging technology. The eye always assumes that the light comes from the straight direction, thus realizing the 3D image on the screen. The solution to reducing the size and weight of the device is basically achieved by reducing the mass of the convex lens with different lens combinations.

AR technology has to be able to image and not block the real object in front of your eyes, which makes the optical structure more complicated, and there are three options for now. The first is called off-axis reflection, which simply means the desired light is reflected back to the eye through a lens with a transparent and reflective surface. The second is birdbath, in which a light source is projected onto a spectroscopic lens that is 45° from the flat edge of the retina. The spectroscopic lens reflects and transmits the light source at the same time. The advantage of this

Fig. 8.2 Optical structure of VR

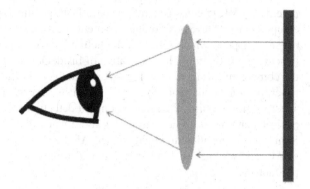

approach is that it has a smaller optical structure than off-axis reflection. The third is optical waveguide, which is more complicated than the first two. Generally, it can be divided into geometric optical waveguide and diffracted optical waveguide. In short, it makes the light constantly conducted by reflection on the inside side of the lens and finally enter the eye, which saves space and makes the whole lens very thin. The optical structure of AR is shown in Fig. 8.3. If only thinness is considered, it is clear that optical waveguides are the most likely final solution for AR technology, but there is another criterion that needs to be considered, namely field of view (FOV). The largest FOV of the current solutions is only 100°, while the thinnest optical waveguide technology can only achieve 50°. The FOV is the included angle between the two edges of the maximum range that the object image of the measured target can pass through the lens. It is called the maximum FOV. For example, when you are playing a 3D shooting game such as CS, the computer screen is partially blocked by the left and the right, which is the FOV reduction.

In addition to the differences in optical structure, AR and VR also have some differences in positioning and recognition technology. In VR, all the scenes are virtual, and only the head position and body posture need to be positioned in real

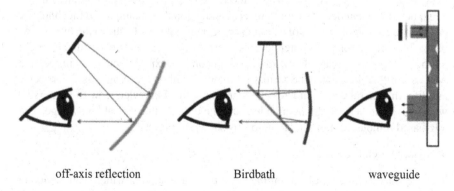

off-axis reflection   Birdbath   waveguide

Fig. 8.3 Optical structure of AR

time. Early VR goggles generally used 6-DOF positioning, which required off-site equipment to achieve positioning. In recent years, InsideOut positioning technology began to appear, which relies on the helmet's own camera to shoot the real scene to achieve positioning. The application of InsideOut technology has greatly reduced the complexity of the device. For AR technology, optical structure can only be the superposition of the real world and the virtual world. However, the accuracy of superposition requires the recognition and detection of various features in the real environment, so as to determine the coordinates of virtual objects in this space. Simultaneous location and mapping (SLAM) technology can be said to be the core of AR, and whether SLAM is rapid and accurate directly determines the quality of products.

8.2   Development of AR/VR Technology

VR/AR technology is mainly applied in the following three areas.

1.   Game and entertainment

Many game companies have launched VR games and supporting equipment, which has subverted the traditional game input mode requiring keyboard and mouse controller, strengthened the interaction between the players and the game scene and enriched the game experience.

2.   Service of life

The application of VR technology in tourism is also very attractive. The analysis and prediction results show that according to the number of domestic and foreign tourists in recent years based on network data, the global travel will reach 1.8 billion by 2030. The application of VR supported by 3D can enable people to see all over the world without leaving home. In the aspect of education and training, the application of AR/VR technology has not only enhanced the interest of people to acquire knowledge, but also expanded the way to receive education. For example, VR technology can realize the establishment of a laboratory in a virtual world. Due to the limitations of teaching conditions and potential risk in experiment, a lot of experiments might not be possible in the past. Now, students can conduct experiments in an immersive way as long as they enter the virtual laboratory, without causing loss of property and life. In general class, students can learn theoretical knowledge in an immersive way, which can better help them understand the obscure theoretical knowledge. The VR-based remote teaching in laboratory is shown in Fig. 8.4.

3.   Commercial service

The US has commercialized military simulation training, which can simulate different battlefield environments, greatly saving military costs on location and unnecessary casualties. In the field of individual training, the training content has

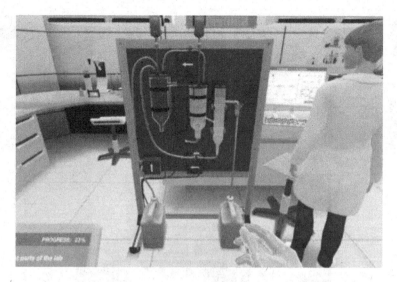

Fig. 8.4 VR-based remote teaching in laboratory

been enriched, and even the joint exercise of multiple services has been realized. In the industrial sector, VR technology has provided a new way to train employees in the operation of equipment. In the virtual world, trainees can familiarize themselves with the environment and equipment of factories before they start working, which will be a great help to the industry in China toward the era of Industry 4.0. In the medical field, the demand for VR/AR technology is particularly prominent. On the one hand, there are very few experts in various fields, and on the other hand, the demand for various consultation is surging. VR/AR technology has provided a solution to this contradiction. In virtual surgery training, for example, with the combination of 3D visualization system and VR technology, remodeling of tissues and organs can be realized in the virtual environment, which is convenient for medical communication and study. The low cost has made VR system play an important role in the cultivation of medical students, especially in the improvement of surgical techniques, so that doctors can make more sophisticated surgical plans and improve the success rate of surgery. AR/VR technology has been widely used in construction, security, aerospace and other fields, and has full space to develop in the future (see Fig. 8.5).

8.3 Applications of VR/AR Technology in Smart Port

VR/AR technology has become more and more popular in games, entertainment, etc., and also been increasingly applied in the industries, so as in the port industry. The application in port is also called port container simulation system based on

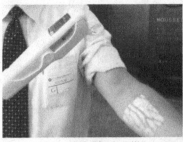
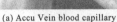

(a) Accu Vein blood capillary   (b) Application of SCOPE AR in site layout

(c) NASA Mars Surface Simulator

Fig. 8.5 Application of VR/AR technology in medical care, construction, aerospace and other industries

VR, which mainly includes virtual training system for port practice base, digital supervision platform for dangerous goods in port, port facility simulation based on VR and so on.

8.3.1 Facility Operation and Business Training in Smart Port

With the effective application of more and more science and technologies in port, many traditional port operations have undergone significant changes in their workflow and employee capacity requirements. For example, in the tally business of the traditional terminals, the tallyman is required to have the relevant ability and master the six elements of tally, namely the container no. of the operating ship, container type, lead seal condition, damage, dangerous goods mark, and the position in ship of the container for loading. He also has to monitor whether the handled containers conform to the actual baplie, verify and confirm the container appearance and seal condition, summarize unloading data of imported containers, record and make complete baplie of exported containers. However, for the smart container ports, intelligent tally system based on OCR technology is being gradually promoted. By means of camera, video monitoring and intelligent recognition, the container tally operation can be completed, which has improved the efficiency of container tally operation in the ports by at least 100%. An obvious example is the intelligent tally system by the QC that has been put into use in the Zhenghe Container Terminals Co. Ltd at Taicang Port. According to statistic data, after the implementation of the intelligent tally system, more than 17 million yuan of labor costs can be saved every year. While the operation efficiency of the tallyman has been greatly improved and the operation cost reduced, it is necessary for the tallyman to have the ability in informatization to operate the intelligent tally system skillfully.

For the traditional tally training, on the one hand, the actual environment is tough, and there are safety risks for trainees in the on-site training and may affect the normal operation of the terminal. On the other hand, the training knowledge explained and illustrated in the classroom is abstract and boring, and the training effect is greatly reduced. For the intelligent tally training, since the number of operation stations has been greatly reduced in the intelligent tally system, it is difficult to organize centralized training. Also, the intelligent tally system is generally in operation around the clock, there is no spare time to assist the staff training tasks in terminal. If the training is forcibly inserted into the operation process, it will bring great safety risk to the operation of the terminal.

In order to solve the conflict of the training, safety risks and operations, VR technology is one of the best solutions. With the help of the tally training system based on VR technology, it can quickly build the working environment of on-site tally or intelligent tally. Through dynamic and random generation of different container data, mechanical failures and disasters, various tally conditions can be simulated, which enables the trainee to quickly master the operation and corresponding counter measures under different working conditions. In this way, it can not only satisfy the requirements of the traditional on-site tally training and intelligent remote tally training, but also make training effectively connected with the actual tally, reducing the training time and costs, and avoiding the potential safety hazard caused by improper operation due to the employee's unfamiliarity with the equipment. The intelligent tally training system is shown in Figs. 8.6, 8.7, 8.8 and 8.9.

Smart port is the inevitable trend of the terminal development in the future, and intelligent tally is an important part of smart port. With the application of intelligent tally technology, visual image and other new technologies, the whole-course monitoring of container operation, real-time comparison of handling information, automatic verification, electronic damage inspection and other work can be realized. These advanced technologies also determine the cost of intelligent tally equipment. Therefore, it is one of the most cost-effective training methods to make a virtual teaching system based on VR technology and train employees for the equipment of intelligent tally system and the work process of on-site tally.

8.3.2 3D Visualization Supervision of Smart Ports

Based on VR technology, IOT and other technologies, the 3D visualization supervision monitoring storage yards, warehouses, berths, containers, ships, equipment

Fig. 8.6 Virtual operation of VR training system of intelligent tally

Fig. 8.7 Tally training with VR training system

and facilities is constructed, with the operation and security of the entire port as the supervision focus. Video monitoring, berth management, yard management, warehouse management, sensor management, emergency response, disaster exercise, big data analysis, decision-making assistance and other systems are integrated to build a 3D visualization platform of port, including displaying, monitoring, warning, positioning and analysis, which achieves comprehensive data integration, intuitive and

Fig. 8.8 Mode switch of intelligent tally system

Fig. 8.9 Operation simulation of traditional on-site tally

visual information, intelligent real-time warning, standardized and efficient disposal, and plays a powerful role in realizing flat and intensive operation of smart ports.

Shanghai Gangcheng Dangerous Goods Logistics Co., Ltd. has realized real-time, visualized and alerted dangerous goods information with the help of 3D monitoring system of dangerous goods storage yard, which has improved the safety supervision level and emergency rescue capability of dangerous goods in port. The 3D system uses VR technology, combined with the AI algorithm, to realize the functions of storage

Fig. 8.10 On-site inspection of dangerous goods container

supervision of dangerous goods containers in the yard, video monitoring, temperature regulation of reefer containers, gas leakage detection, electronic perimeter regulation, wind speed and direction regulation, dangerous goods warehouse supervision, building mark management, risk assessment and disaster response, fire drill, etc. The system can assist the supervision personnel of the company to manage, monitor and rescue the dangerous goods containers (Fig. 8.10) in the storage yard more accurately, efficiently and scientifically.

Benefiting from VR technology, the system mainly has the following characteristics.

1.   3D models close to the real

Modeling and 3D rendering of the dangerous goods containers in the yard, such as general containers, tank containers, reefer containers, high and low containers, as well as unpacked dangerous goods in the yard and warehouse and other regulatory objects have been achieved. Meanwhile, buildings, equipment and facilities, roads and terrain environment in the yard have been restored and displayed in equal proportions, to provide a realistic virtual regulatory environment for the supervisors of relevant departments and build a foundation for the operation of VR platform in the future. As shown in Fig. 8.11, with the VR platform, the operators can view the real-time information of each container in the scene, such as the container no., container location, container type, dangerous goods type, UN no., weight, arrival time, storage days, cargo description, port zone and dangerous goods nature, etc. In addition, the operators can also fulfill quick container query, statistics, positioning and other practical functions on the VR platform, see Fig. 8.12.

Fig. 8.11   3D supervision panorama of dangerous goods container yard

Fig. 8.12   Information query of dangerous goods containers

2.   Multi-dimensional real-time monitoring

The monitoring function of the VR platform is realized through connecting to a large number of monitoring sensors or alarm systems in the port, including multiple monitoring cameras and NVR equipment, anemometers, temperature sensors of reefer containers, hazardous gas sensors, perimeter alarm system, etc. Meanwhile, with

Fig. 8.13 Data monitoring of hazardous gas sensors

the communication and database technology, the system can synchronously update these regulatory data and response and process correspondingly. Data supervision of hazardous gas sensors is shown in Fig. 8.13.

3.　Professional message alerts

The system has a professional knowledge base for the supervision of dangerous goods containers, covering all 9 hazardous classes and all UN dangerous goods no. (UN no.). It can not only retrieve and query the location and relevant information of containers according to the specified hazardous classes or UN no., but also generate corresponding disposal and emergency response information for these hazardous classes or UN no. in real time and display them directly on the interface to assist supervisors to complete disposal and response operations. Interactive between virtual simulation and security camera is shown in Fig. 8.14.

4.　Friendly interface

In addition to presenting intuitive 3D scene information, the system also simplifies a lot of complicated operation procedures according to the characteristics of 3D presentation, so that the supervisory operators can complete the supervisory operations quickly and conveniently. Moreover, besides the traditional mouse and keyboard to complete the entire operation process, the external equipment such as rocker and handle can also be applied to complete operation.

The development and application of 3D supervisory platform in smart port, can effectively integrate and configure the basic data resources of container and cargo in the port, so as to realize the digitalization and visualization of fundamental data

Fig. 8.14 Interactive between virtual simulation and security camera

resources management, further promote the construction of geographic information service platform, realize the connection and sharing of data resources between
administrative staff and operators in the port. Meanwhile, the VR platform can also
realize the simulation of emergencies and the practice of the ability to deal with
emergencies. Thus, the supervision level and efficiency of the terminal company
on dangerous goods containers are improved, the professional requirements for the
supervision personnel are reduced, and the human cost and economic cost of supervision are reduced. At the same time, it ensures the smooth progress of port safety
supervision. Fire drills and emergency disposal carried out through the VR platform
are shown in Fig. 8.15.

8.3.3 Interactive Simulation of Machinery Equipment in Smart Port

In recent years, with the progress of computer simulation software, computer is
applied to simulate the new technologies, new plans and new facilities proposed in
the smart port, which provides reference for the decision makers of the smart port
and the designers of automatic port machinery equipment. Simulation technology is
playing a significant role in port planning, port operation plan design, port operation
scheduling, port machinery equipment testing and so on. For instance, for the research
and debugging of system equipment in automated terminal, the software of each
project completed by designers can be run in the office using the simulation platform
to test the accuracy, integrity, stability and other performance of the software, so
as to reduce or even avoid on-site testing and modification. The software tested by
the simulation platform covers equipment control system (ECS) (including QCMS,
BMS, VMS, etc.), automatic crane control system (including QC ACCS, ARMG

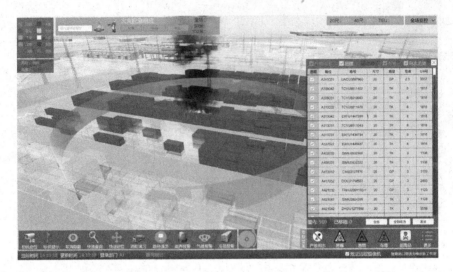

Fig. 8.15 Fire drills and emergency disposal

ACCS, AGV ACCS, etc.), crane PLC program (QC, ARMG, AGV, etc.). The remote operation station in port is shown in Fig. 8.16.

Completely based on the simulation software development process, however, only the verification of system structure, principle and algorithm are realized, and the performance of the final prototype hardware system is difficult to be guaranteed. On the one hand, there is no simulation test on the hardware of the system, and on the other hand, there are often problems such as the unreliability in software code or

Fig. 8.16 Remote operation station in port

even in the hardware environment for code running, which eventually leads to the increase of project cycle and cost, and may even lead to the failure of the project.

Therefore, it is necessary to combine software simulation with real hardware controllers (sensors, PLC controllers, remote control station, etc.) to conduct interactive simulation. After ensuring the necessary hardware or software environment in some actual operations, the participation ratio of simulation software should be increased. Thus, while ensuring the accuracy of the test results, the test cost and test cycle of the automatic port machinery equipment are reduced to the maximum extent (Fig. 8.17).

With the help of VR simulation system of stacker-reclaimers, a well-known automatic port machine enterprise in China has realized the interactive operation debugging of remote-control platform and simulation platform of stacker-reclaimer. The remote-control platform of the stacker-reclaimer is installed in the central control room. It can communicate with any of the four stacker-reclaimers in the VR simulation platform through the Modbus protocol and control it remotely, meanwhile, the other three are automatically controlled through the AI module of the VR simulation platform. The AI module of the remote operation platform and the simulation platform can realize the cooperative and interactive control of all the stacker-reclaimers in the yard. Thus, a simulation environment for the automatic control system and the remote-control system of the stacker-reclaimers is provided, which can be designed, developed and debugged in the laboratory, and the resource waste and safety risk of the on-site prototype experiment are avoided.

Meanwhile, when the port machine or its component mechanism that the system needs to control changes, it only needs to modify and adjust the virtual model of port

Fig. 8.17 VR interactive simulation of stacker-reclaimers

Fig. 8.18   The virtual remote cameras

machine in the 3D virtual scene to rebuild the physical model and dynamic model, and
then use VR technology to carry out simulation experiments. The operator only needs
to control the virtual platform in the operation room to get the dynamic performance
of the stacker-reclaimer in the operation process of the terminal, which provides a
direction for improving the R&D efficiency and reducing the cost of the product.

The VR simulation system of the stacker-reclaimers has the following characteristics.

1. Virtual remote cameras (Fig. 8.18). With this technology, multiple monitoring
   areas on the same screen can be tracked, a camera monitored area can be focused
   on amplification, and the panoramic camera can be switched, etc. These advantages make the VR system more able to truly reflect the reality and facilitate the
   observation of experiments.
2. Multiple controlling. By switching the control button, handle and other functional buttons of the remote-control station, all mechanical facilities in the same
   virtual scene can be controlled, and the interaction and cooperation of multiple
   control stations to all mechanical facilities in the same virtual scene can also be
   realized.
3. Rich virtual sensors, such as virtual radar, virtual limit sensor, virtual encoder,
   virtual flow scale, virtual fault, etc. The condition information of all virtual
   sensors can be fed back to the main control PLC at the background for its
   implementation of the corresponding logic. Adding a variety of virtual sensors
   is conducive to more accurate model of the real on-site operation conditions.
4. Real kinematic models and physical properties. The kinematics models include
   the action of the main mechanism, the action of the conveyor belt, and the
   change of the shape of the material pile when the material is taken and thrown,
   etc. Physical effects include gravity, resistance, inertia, friction, as well as the
   free fall, collision, overturning, rolling, vibration and other physical effects of

Fig. 8.19 The dynamic generation of the material pile

    mechanism, material pile particles and other goods. The dynamic generation of the material pile is shown in Fig. 8.19.

5. Anti-collision detection. The collision detection areas include 3D sphere, 3D cube and other irregular areas. All entities can be the objects of collision detection, and the layered detection logic of each entity does not interfere with each other. The virtual sensor collision detection of the bucket-wheel of the stacker-reclaimer is shown in Fig. 8.20.

6. Rich communication protocols, including OPC Server, Modbus, TCP/UDP, Profibus and other programmable general standard protocols. With these protocols, functions such as the main control PLC's real-time control of all the mechanical facilities of the virtual remote station as well as the real-time feedback of the sensor information and condition information of each virtual mechanical facility to the main control PLC, can be realized.

8.3.4 AR Technology Serving Smart Port

AR is a new human–computer interaction technology. The scene that people see is presented after technical processing by AR equipment. It is a superimposed digital image based on the real environment, and also has some motion tracking and feedback technology. The application of AR technology in smart port is still in a small scope at the present stage. Thanks to the wide use and rapid development of camera and intelligent identification technology, AR technology will be widely used in all aspects of smart port in the future.

    The operating environment in the port is harsh, and all kinds of port machinery equipment are in continuous high-load operation condition most of the time, and the probability of mechanical failures increases accordingly. Therefore, the maintenance

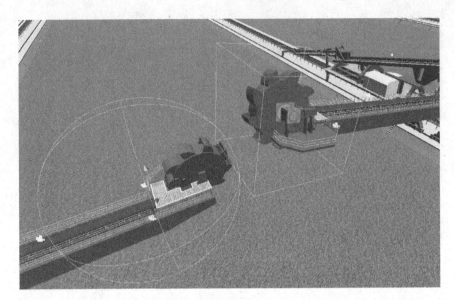

Fig. 8.20 The virtual sensor collision detection of the bucket-wheel of the stacker-reclaimer

expense and duration of port machinery equipment are the costs that must be considered by the terminal company. Port machinery are complex with multiple types, and now the intelligent process of terminal is accelerated, and the upgrading of equipment often takes place, which brings great challenges to maintenance personnel. In this case, even the most experienced maintenance personnel will inevitably meet situations that the cause of the failure cannot be very clearly determined and fast maintenance is impossible. In view of this situation, the introduction of AR technology in the maintenance process of port machinery equipment can obviously shorten the maintenance cycle.

The technology allows maintenance personnel to recognize the unique identification code of the equipment in a database through a phone camera or smart glasses, and quickly retrieve information about the equipment's operating parameters, maintenance history, internal structure (Fig. 8.21), cause of failure and treatment and other information needed by maintenance personnel. In order to realize the display of virtual equipment information and real equipment scene on the same screen, the registration and matching of virtual equipment information and real equipment are required at the location in 3D space. This is mainly realized by tracking technology, using mobile camera to detect feature points and contours of equipment, tracking feature points of the object to automatically generate 2D or 3D coordinate information and matching with the location of the virtual information, and finally displaying the possible failure and treatment at the location by retrieving equipment information from the database.

With this technology, the range of possible failures is greatly narrowed. In this case, maintenance personnel only need to combine the data provided by AR and

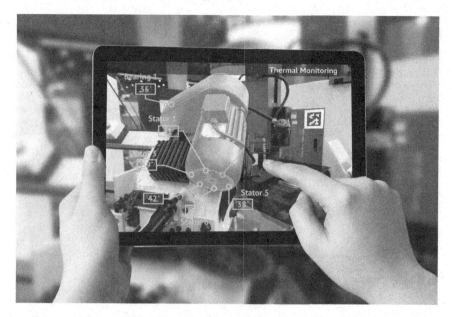

Fig. 8.21 Checking the internal structure of the equipment with AR technology

their own maintenance experience to determine the cause of the failure and the maintenance method (Fig. 8.22). This can greatly reduce the operation intensity, for example, the maintenance of equipment does not need to search for maintenance

Fig. 8.22 Equipment failure treatment with AR technology

Fig. 8.23 AR tracking navigation of Amap

records and fault diagnosis instructions page by page, thus improving the operation efficiency.

Another application of AR technology is navigation system. As an innovative interactive way, this technology brings new ideas to map navigation. Unlike the traditional map navigation, vehicle-mounted AR navigation first using camera to real-timely capture the real scene of the road ahead, then conducting fusion calculation coupled with the current location, map navigation information and AI recognition of scene, then generating the virtual navigation guidance model and adding to the real route, to create navigation screen closer to the driver's real vision, which has greatly reduced the user's cost of using traditional 2D or 3D electronic maps. AR tracking navigation of Amap is shown in Fig. 8.23.

This technology can also be applied in smart ports to provide tracking, identification and navigation capabilities for various handling and transportation facilities. For example, for the external container trucks in the port, the operation route of the terminal is complex and the driver does not know all the road information, which often leads to the occurrence of accidents. In order to avoid traffic congestion and other accidents caused by drivers' failure to obtain the road information of the terminal in time, GPS/Beidou and AR technology are combined to improve the safety of container truck navigation, and the navigation information is more comprehensive. This is a holographic AR navigation system, which will constantly update the map information to provide navigation as the surrounding environment of the truck changes, and automatically select the optimal route in the road section where the accident occurs. Drivers do not need to be equipped with a headset to get a vivid, accurate holograms. This technology can accurately display the direction of the vehicle according to the speed of the vehicle, and the navigation indicator arrow is projected onto the road via the display mounted on the instrument panel, allowing the driver to use the navigation without being distracted, thus achieving safe driving. By introducing AR navigation module, the port can be more intelligent and the safety and efficiency of operation can be greatly improved.

In general, VR/AR technology has been effectively applied in many fields such as retail, construction, tourism, education, medical care, military, leisure and entertainment in China, and also has great development potential in the port and shipping industry. The development of the industry needs to be driven by business. Since 2016, Google, Apple, Facebook, Lenovo and other technology giants have invested and planned in this field, mainly providing consumers with immersive VR/AR devices and AR/VR solutions needed in the development process, which has fully shown that VR/AR technology has a huge market and space for development. At present, the development of VR/AR technology in China is still at the beginning stage. It is believed that in the near future, VR/AR technology will further penetrate into various industries in China and play a pivotal role.

Bibliography

1. Qi H, Wang X (2004) System modelling and simulating. Tsinghua University Press, Beijing
2. Zhang X (2005) Digital simulation of control system and CAD. China Machine Press, Beijing
3. Zhao N, Xia MJ, Xu ZQ et al (2015) A cloud computing-based college-enterprise classroom training method. World Trans Eng Technol Educ 13(1):116–120
4. Zhao N, Xia M, Mi C et al (2015) Simulation-based optimization for storage allocation problem of outbound containers in automated container terminals. Math Probl Eng 2015
5. Zhao N, Shen Y, Mi C et al (2015) Vehicle-mounted task control system in container yard based on workflow engine. J Coast Res 73(10073):220–227

Chapter 9
Smart Port and System Simulation/Emulation

9.1 Concept of System Simulation

In the known world of mankind, from the vast universe to the microscopic world inside an atom, all can be described with systems. As shown in Fig. 9.1, in the system theory, astronomical system and microscopic system can usually be defined as quantum systems; society, economy, ecological system, hydromechatronics system can be summarized into the research category of continuous system; military, industry, transportation and logistics are usually discrete systems. System simulation is to establish a simulation model which can describe the system structure or behavior process and has certain logical or quantitative relationships, based on the analysis of properties of all elements in the system and their mutual relationship, according to the purpose of system analysis, to conduct experimental and quantitative analysis, in order to obtain all kinds of information needed for the correct decisions.

Port logistics system simulation mainly studies the organization, operation and management of port logistics, which belongs to discrete event simulation (DES). Distributed system usually contains the following elements: entities, attributes, activities, events, and state variables. Entities are the main objects and components in the system, attributes describe the characteristics of entities, activities describe subject behaviors, events are immediate situations that can change the state of a system, state variable represents a collection of variables in the system. System simulation of port distributed events, is by studying all kinds of logistics entities in the discrete system, such as quay cranes, yard cranes, container trucks, AGVs in the process of operation of various kinds of attributes, activities, events, and states, to establish DES model, on the basis of experimental or quantitative analysis, to obtain the correct decisions required for all kinds of information. For example, the port logistics system can be simulated to analyze the port logistics system capability, system bottleneck, rationality of resource allocation, advantages and disadvantages of management decision-making methods, etc.

The system simulation in port also includes the continuous simulation of the hydromechatronics systems of large port machinery equipment. The most common

© Shanghai Scientific and Technical Publishers and Springer Nature Singapore Pte Ltd. 2022
W. Mi and Y. Liu, *Smart Ports*, https://doi.org/10.1007/978-981-16-9889-7_9

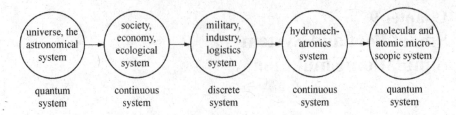

Fig. 9.1 System types

applications are analysis and optimization in mechanical design of large port machinery equipment with the finite element simulation technology, and simulation analysis and optimization of electric control system, hydraulic system and other large equipment control system with Siemens, Dassault and other professional engineering simulation software.

In the era of Industry 4.0, with the development of AI technology, in the foreseeable future, due to the strong ability in time and space expansion inherent in simulation, system simulation technology will be integrated with AI technology, deeply affecting the future development of military, industry, logistics and other fields.

9.2 Development of System Simulation

The main discrete event system simulation software was developed from the end of 1980s to the first decade of twenty-first century, as shown in Table 9.1. From the end of 1980s to 2006 was a peak period for the development of commercial simulation software. During this period, on the one hand, accompanied by the industrial globalization, developed countries had turned away from heavy industrial production to higher value-added service industry. On the other hand, rapid advances in computer technology had made complex system simulation software popular. After 2010, the fourth industrial revolution (intelligent manufacturing) came, and industrial software giants represented by Siemens and Dassault began to develop intelligent manufacturing platforms from product design to maintenance of the whole life cycle, and continued to acquire and integrate basic industrial software in all sections of the whole industrial chain. Discrete event system simulation software, also faced a huge impact in this period, either acquired or began to focus on some unique areas.

It is worth noting that in recent years, with the development of new technologies such as computer technology, network communication and industrial intelligence, some research teams have also been constantly exploring and developing DES packages adapted to recent technological developments. For example, SimGe, an automated modeling tool based on high-level architecture (HLA) developed by Okan Topçu team, can quickly realize HLA modeling and integration of different structure systems such as control, management and simulation with distributed simulation. Ucar team, on the other hand, has developed a DES package that can use R

Table 9.1 Development of the main commercial DES software

Software	Development time	Development company
ExtendSim	1988	ImagineThat Inc
Plant Simulation	1992	Simple + +, Adsop/Technomatix/KG, later acquired by Siemens
Arena	1993	System modeling company
Simul8	1995	SIMUL8 company
AutoMod	1996	AutoSimulations Inc
ProModel	1996	ProModel corporation
FlexSim	1998	F&H, later acquired by Flexsim
Anylogic	1998	DCN research group of St. Petersburg Technical University
Witness	2001	Lanner group
MicroSaint	2003	Micro analysis and design company
Quest	2003	Delmia corp, later acquired by Dassault group
Simio	2006	Simio LLC company

language, which is suitable for developers familiar with R language. The intelligent digital twin platform developed by National University of Singapore and Shanghai Maritime University based on the.net DES framework O2DES is committed to the combination of DES and the latest network, intelligent and visualized technologies to realize the application of intelligent digital twin system. At present, important progress has been made in the integration of TOS, ECS and 3D real-time simulation in the automated container terminal.

In the application, the initial simulation technology was used as an auxiliary tool for testing the actual system, and then used for training purposes. Now the application of the simulation system involve in aerospace, aviation, research and development of various weapon systems, electric power, transportation, communication, chemical industry, nuclear energy, automobile, shipbuilding and other industries, mainly used for system concept research, system feasibility study, system analysis and design, system development, system testing and evaluation, system operator training, system forecasting, system utilization and maintenance and other aspects.

1. Applications in military

The initial application of DES technology in military field is military training and war game deduction. Distributed simulation system connects people scattered in different places, simulators in the loop, armed forces generated by computer and so on into a whole through the Internet, forming a virtual battlefield that can be coupled in time and space. With distributed virtual battlefield simulation, strategic and tactical drills and training which are difficult to be carried out effectively in a peaceful environment are realized. Secondly, the application is involved in the research and development of new weapon systems, especially after the maturity of modern simulation technology, the distributed simulation system can realize the seamless connection between DES and continuous simulation, and realize the comprehensive simulation application in multiple physical fields. Many complex weapons systems, such as aircraft carriers, submarines, fighter jets, and hypersonic weapons, cannot be developed without modern simulation technology. Thirdly, it is the research of weapon production system, military logistics support system and so on. The simulation system provides scientific management services for the production, deployment, transportation, storage, daily maintenance and other sections of military logistics supporting materials such as weapons and equipment. Finally, DES can also provide analysis and support for the deployment of social emergency events requiring military support, such as rescue and disaster relief.

2.   Applications in industry

Due to the complexity of industrial systems, from the perspective of security and economy, DES technology has been widely used in various sectors in the industrial field, and plays an important role in the construction plan research of large-scale complex engineering systems and the system operation management process.

DES has penetrated into the following aspects in the automobile manufacturing process, such as design plan demonstration of factory and manufacturing assembly line, production planning of auto manufacturing process, logistics distribution of production line, design and management of supply chain system of thousands of auto components, design and management of global automobile sales and logistics network.

In marine engineering, port machinery, shipbuilding and other large equipment industries, DES technology is also playing an increasingly important role in production design, material management in manufacturing process, component hoisting and assembly, product handling process, logistics engineering in shipping and land transportation, etc.

In the electric power industry, with the increasing capacity of the unit generator and the increasing complexity of the system, higher and higher requirements are put forward for its economic operation and safe production. DES can play an increasingly important role in the construction plan and operation management of power stations. Meanwhile, DES technology can also play an important role in the study of the complex relationship between power allocation, social production and urban life in the deployment process of complex and smart national grid.

3.   Applications in education and training

In many complex systems with certain economic or security risks, personnel operation and management are indispensable. Such system faces great difficulties in the education and training of special operation and management personnel, such as the operation and management of nuclear power plant, engine test in aerospace industry, large ship steering, operation of large port machines and so on.

Similar to the research and development of advanced weapon system, the distributed simulation system combines DES with continuous simulation to realize the integrated application of multi-physical field simulation. Combined with advanced VR technology, the immersive and realistic 3D virtual simulation application of complex system can be realized. These simulation applications have great value for solving the problems of personnel operation of complex system, management training and education with certain economic or security risks. Through the virtual environment operation and management training, the new employees can get familiar with and experience the special working environment in advance. At a low cost, zero-based students or employees can complete the psychological and operational management of technical adaptation to the special work environment or driving equipment.

The comprehensive simulation application of multi-physical fields is also playing an increasingly important role in basic education and teaching, such as youth science popularization, complex biochemical reaction process, complex virtual geometry and physics.

4. Applications in other fields

In the twenty-first century, discrete event system simulation, especially distributed multi-physical integrated simulation, has gradually penetrated into many fields such as medical treatment, communication, society, economy, entertainment, etc. In recent years, the operation and management of hospitals have become a hot research topic in this field. For example, the construction of a random group movement model can effectively simulate various queuing and service systems in large hospitals. The rational allocation of medical resources in hospitals is also one of the focuses of attention. Traffic network planning and traffic control simulation have always been important research directions of DES technology in transportation industry. For port logistics, DES is an indispensable and important research means.

9.3 Applications of System Simulation in Smart Port

Discrete event system simulation has been widely used in port logistics, such as research in sea and land traffic network system, comprehensive simulation of large-scale port equipment, training of port equipment operators, evaluation and optimization of port operation management system, etc. This section will take the applications of discrete event system simulation in plan and design of container terminal as an example to introduce the applications of discrete event system simulation in port. These applications are mainly derived from the design and simulation of ultra-large scale container terminals in the Next Generation Port Challenge in Singapore, the design simulation of automated container terminal in Haifa, Israel, and automated container terminal in Abu Dhabi, United Arab Emirates.

The Next Generation Port Challenge in Singapore aimed to design an automated container terminal with an annual throughput of 20 million TEU in a rectangular area with a length of 2.5 km, a width of 1 km, three sides adjacent to the sea, and a width

of 1 km connecting to the land. The regional schematic diagram and final design
are shown in Figs. 9.2 and 9.3. The outstanding features of the final design are as
follows. First, the capacity of container storage and horizontal transportation of the
terminal has been fully improved with the double-deck structure. Second, double-
deck handling at the quay has been achieved with the three-trolley quay cranes, the
unit throughput capacity along the shoreline has been broken through, and the ship
loading and unloading operations have been speeded up. Third, energy saving and
emission reduction of large-scale equipment cluster have been achieved with smart
grid, solar energy and other green energy systems, as well as ARMGs equipped with
weight balancing device. Fourth, the logistics collection and distribution capacity of
the port area has been greatly improved by the integration of large-scale logistics
centers. System simulation technology plays an important role in the optimization

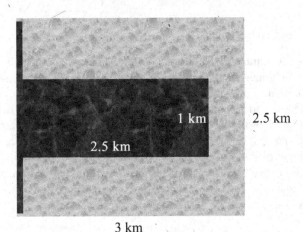

Fig. 9.2 The regional schematic diagram of the Next Generation Port Challenge in Singapore

Fig. 9.3 Final design of the Next Generation Port Challenge in Singapore

and verification in the overall throughput capacity of the final design, the quayside loading and unloading system, the horizontal transportation system, the yard loading and unloading system, the gate distribution system, etc. For example, in this case, with the system simulation technology, it is demonstrated that the proposed design will not only meet the annual throughput capacity of more than 20 million TEU, but also can achieve the on-time berthing rate of 93% ships (within 2 h) and land productivity of 1,768 TEU/ ha./hour under the given ship arrival. It is 185% higher than the Euromax terminal in Rotterdam and 321% higher than the ETC DETA.

The simulation of the design plan of the automated container terminal in Haifa, Israel is entrusted by China Communications Construction Company (CCCC) Third Harbor Consultants Co., Ltd., to carry out the capability simulation demonstration for the design plan. The main research objectives include the following aspects, to study the quay loading and unloading capacity supported by the overall operation of the terminal and the annual throughput capacity of the terminal under the existing design plan; to study the horizontal transportation capacity of the terminal and the pressure of the main roads and traffic intersections under the design plan; to study the loading and unloading capacity of the container storage area; to study the passing capacity of the gate and the queuing situation of vehicles at each gate. The design plan of the automated container terminal of Haifa Port in Israel is shown in Fig. 9.4.

The Port of Abu Dhabi is located on a small island off the central coast of the United Arab Emirates, connected to the land by bridges and seawall, and adjacent to Mina Zayed to the east, which mainly serves Abu Dhabi. Abu Dhabi's industry is dominated by petrochemicals, natural liquefied gas, aluminum metallurgy, plastic products, clothing and food processing. Abu Dhabi is also a tourist destination with modern factories, convenient transportation, and thriving businesses. The simulation analysis study on the design plan of the automated container terminal of Abu Dhabi Port is similar to that of the automated container terminal of Haifa Port in Israel, which demonstrates the overall service capability of the terminal on the basis of a

Fig. 9.4 The design plan of the automated container terminal of Haifa Port in Israel

given planning and design. The simulation analysis includes the overall throughput capacity of the terminal, the road network capacity of the horizontal transportation, the operation capacity under the configuration plan of the horizontal transportation equipment, the loading and unloading capacity under the configuration plan of the container storage area, the passing capacity of the gate and the queuing situation of vehicles at each gate.

The following part of this section will introduce the applications of system simulation in port based on these simulation cases of automated container terminal planning and design.

1.   Research on the overall capacity of the terminal

The simulation analysis of terminal's overall capacity is mainly based on the assumption of different facility configurations (quay cranes, yard cranes, horizontal transportation facilities), to simulate the entire operation of the terminal in the comprehensive consideration of the existing design plan, so as to evaluate the comprehensive handling efficiency of quay cranes that the terminal can actually provide, and then evaluate the annual handling capacity of the terminal. By analyzing the investment benefit of different layout, different facility configuration plans and handling capacity of the terminal, it provides scientific decision-making reference for decision makers. Figure 9.5 shows the throughput capacity of a terminal with different facility configurations. Figure 9.6 shows the payback years of the increased facility investment

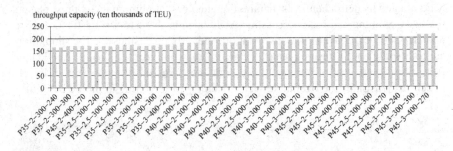

Fig. 9.5 The throughput capacity of a terminal with different facility configurations

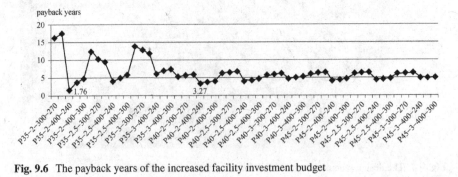

Fig. 9.6 The payback years of the increased facility investment budget

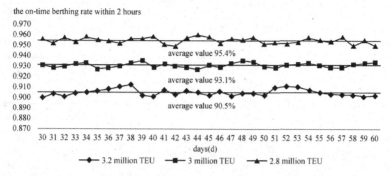

Fig. 9.7 On-time berthing rate under different ship arrival distribution

budget corresponding to the facility configuration plan. The combination of simulation analysis and investment income analysis can provide scientific decision-making reference for decision makers to select the best container design plan.

2. Research on berthing service capacity

The research on berthing service capacity is to study the capacity and service quality of berthing service system under the assumption of different vessel arrival distribution after obtaining a more reasonable comprehensive operation efficiency of quay cranes. The research focus is to investigate the on-time berthing rate and berth utilization rate under different ship arrival rules. Figure 9.7 shows the probability that the waiting time of a ship in the anchorage is no more than 2 h under different ship arrival rules (which can be converted into different annual throughput of the terminal). The x-axis represents the statistical days intercepted in the simulation, and the y-axis represents the on-time berthing rate within 2 h. This index fully reflects the service quality of the terminal.

Figure 9.8 shows the berth utilization obtained through simulation analysis. The x-axis represents the length of terminal shoreline occupied or docked with ships, the left y-axis represents the proportion of shoreline occupied, and the right y-axis represents the cumulative probability. Wherein, occupancy rate refers to the time proportion that the shoreline is docked with ships or other plans, and berthing rate refers to the time proportion that the shoreline is docked with ships. This index can effectively evaluate the utilization of berth resources. If all the shorelines are occupied with a high rate, it indicates that the service capacity of terminal berths is insufficient under the design plan.

3. Research on quay crane utilization rate

Similar to the berthing service capacity study, the quay crane utilization research is to study the utilization of the quay cranes after obtaining a reasonable comprehensive operation efficiency of the quay cranes. Quay crane utilization rate is shown in Fig. 9.9. The x-axis represents the number of quay cranes occupied or under operation, the left y-axis represents the probability of such situation, and the right y-axis

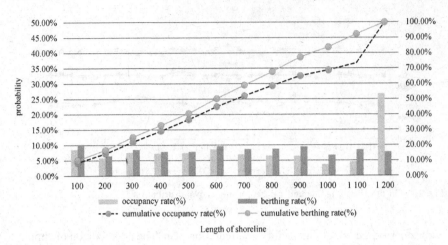

Fig. 9.8 The berth utilization rate

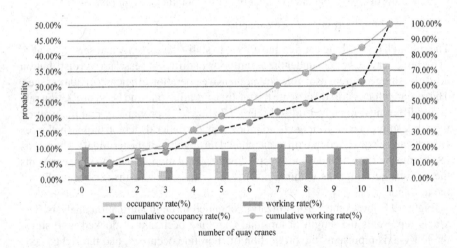

Fig. 9.9 The utilization rate of quay cranes

represents the cumulative probability. A high occupancy rate of 11 quay cranes indicates insufficient quay cranes. When the decision maker needs to increase or decrease the number of configured quay cranes, this index can provide a reliable reference for decision maker.

4.   Research on horizontal transportation system

Research on horizontal transportation system can study the operation efficiency of different horizontal transportation facility configuration systems under selected layout, quay crane and yard crane configuration plan. It is also possible to study the operation of horizontal transportation system under different operation intensities in the condition of determining the complete design plan and facility configuration

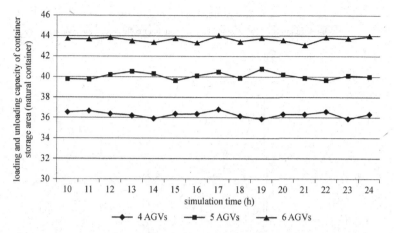

Fig. 9.10 Average operation volume of each operation line under different configuration plans in simulation (container/hour)

plan. Figure 9.10 shows the average efficiency of each operation line with different AGV configurations, which can provide reasonable number of configured AGVs. Figures 9.11 and 9.12 respectively show the average operation time and driving distance of transportation facilities under a given design plan. These indicators can effectively reflect the operation conditions of the horizontal transportation system under the design plan.

5. Research on traffic network system

Traffic network system research mainly studies the theoretical capacity and simulated traffic flow of each road and intersection as well as the relationship between the two under design plan. Through the analysis of the traffic flow of each road and intersection in the simulation, it can effectively evaluate whether the traffic network configuration of the design is reasonable. Meanwhile, through the relationship between the traffic flow and the design capacity (traffic saturation degree) in the simulation, the weak points of the traffic network can be found, which provides guidance and suggestions for the improvement of the traffic network design.

Figure 9.13 shows the graphical display of traffic saturation degree of a terminal traffic network. Different colors in the figure represent different traffic flows. For example, the part circled by the dotted line, with darker colors, indicates that the traffic flow exceeds the designed capacity.

6. Research on the operation capacity of the container storage area

The research on the operation capacity of the container storage area is mainly to study the loading and unloading capacity at the sea end and land end of container storage area (automated container terminal) under the given layout and facility configuration plan, as shown in Fig. 9.14, which can effectively evaluate whether the yard crane configuration capacity of the design plan is reasonable. For the manual terminal, the operation capacity of the container storage area can be increased or decreased by

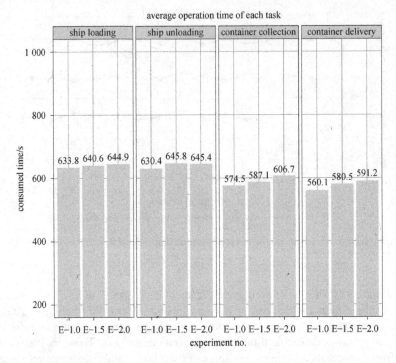

Fig. 9.11 Average operation time of the four basic tasks

adjusting the quantity and equipment parameters of the yard cranes. For the automated container terminal, since RTGs are commonly applied, the different operation
requirements of sea end and land end can be met mainly by adjusting the equipment
parameters of the yard cranes. Since the container storage area of a container terminal
is the core to undertake loading and unloading at sea end as well as collection and
distribution at land end, its loading and unloading operation is complex, and it is the
core area of terminal operation, which determines the actual handling and turnover
capacity of a terminal. Therefore, it is very important to study the operation capacity
of container storage area of the terminal with simulation.

7.   Research on gate service capacity

The research on the service capacity of the gate usually studies queuing situation at the
gate and detention time of the external container trucks at the port based on the given
design plan, under the assumption of different arrival distribution rules of the external
container trucks. By observing the queuing situation of external container trucks at the
gate (Fig. 9.15), the adequacy of the capacity of the gate can be effectively assessed.
By observing the time distribution of external container trucks in port, as shown in
Fig. 9.16, the rationality of the distribution of terminal collection and distribution
system can be effectively assessed. Meanwhile, by investigating the operation time
of the external container trucks in the port, it can also reflect the collection and
distribution capacity of the container storage area.

average no-load driving distance when completing a task

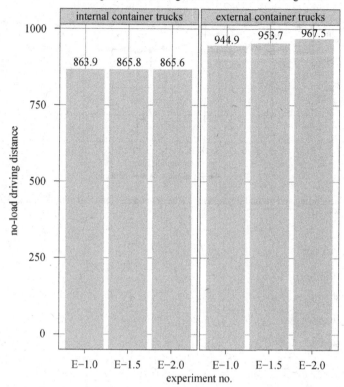

Fig. 9.12 Average driving distance

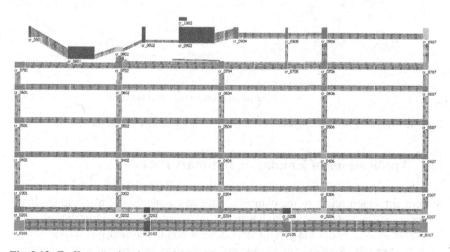

Fig. 9.13 Traffic saturation degree of the traffic network system of the design plan

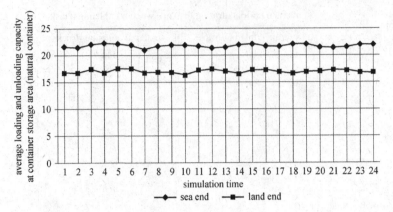

Fig. 9.14 The loading and unloading capacity at the sea end and land end of the simulated container storage area (24 h)

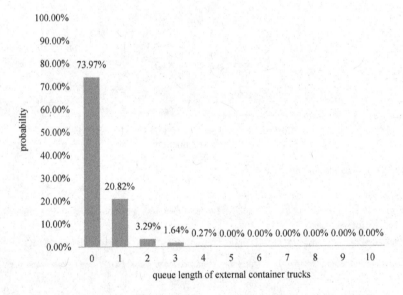

Fig. 9.15 The simulated queuing situation of the external container trucks at the gate

9.4 Applications of Emulation in Smart Port

System emulation refers to the computer simulation analysis of a specific system behavior that has not yet occurred. It is to observe the change of the specific behavior inside the system under the given external input conditions and how it will affect the output results. System emulation is mainly used to analyze and solve a specific business operation problem and examine the roles of many factors that constitute this behavior in the behavior process. The use of computer software to emulate or

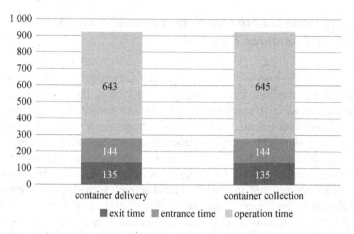

Fig. 9.16 The simulated time distribution of external container trucks in port

represent the behavior and processes of a physical or abstract system is similar to system simulation in this aspect. System simulation applied to simulate the function of a system with computer model, only requires that with the same input, the output of the system simulation and the actual output of the simulated system are consistent, instead of requiring to reflect or realize the internal details of the system behavior. For example, to analyze and simulate the stacking and loading capacity of a container yard, the system simulation will give its overall indicators such as annual container passing capacity, peak container passing capacity and so on. However, the system emulation is to use the computer model to emulate and express the composition and function of each component in a certain system, and to emulate the operation mechanism of the system. This requires that the designer of the system emulation software needs to understand the internal structure and operation mechanism of the emulated system very well, and can use various data structures to realize the model of each component. For instance, in the analysis of the stacking and loading capacity of a container yard, the system emulation will analyze and emulate the passing capacity of the container yard for one ship loading and unloading, as well as the daily, weekly and monthly dynamic passing capacity of the storage yard with the specific collection and distribution sub-model, storage sub-model and facility scheduling sub-model. Obviously, the system emulation analysis results are better matched with the specific operation conditions of each component of the terminal, and have greater guiding value for the optimization and improvement of the operation process. Under certain conditions, these emulation models can be directly embedded into the terminal operation management system. It can also be said that system simulation is to macroscopically analyze the functions and capabilities of the system on the whole, while system emulation is to deduce the functions and capabilities of the system from the actual activity process of the constituent substructures of the system.

9.4.1 Architecture of the Operation Emulation System of the Container Terminal

In order to realize the system emulation of actual activity process of each substructure to deduce the function and the ability of the system, the operation emulation system of the container terminal needs to be able to emulate each small activities of the actual container terminal system, which means the architecture of operation emulation system should be similar to the actual terminal operation system, to reduce the emulation deviation caused by heterogeneity and facilitate the integration and implementation of the system. In addition, as one of the major functions of the emulation is to evaluate and analyze the actual activities, functions and capabilities of each subsystem, the operation emulation system of the container terminal needs to save and present various process data in the emulation process, which requires design of special data structures and interfaces for data statistics. In consideration of characteristics above, the architecture of the operation emulation system of the container terminal is shown in Table 9.2.

In the actual emulation process, in order to get closer to the operation of the actual system and simulate the interaction and operation of various components of the system in actual operation activities, the system structure combined with multi-agent as shown in Fig. 9.17 is adopted in the operation emulation system of container terminal.

In actual operation activities, various operation tasks in the container terminal usually require the same or more types of facilities to cooperate with planners and schedulers to complete. In the emulation system, the agents act as cooperative structures of facilities and operators with autonomous decision-making ability. For example, the container truck agent is equivalent to the container truck operated by the driver, which not only realizes the function of the container truck to execute operation instructions, but also simulates the micro decisions of the driver in the driving process, such as avoiding and deadlock processing. If the scenario is changed to an automated terminal with the intelligent container truck as the main horizontal transportation facility, similarly, the intelligent container truck agent simulates the processing of the intelligent container truck and also has the avoiding and deadlock processing functions of the intelligent container truck. Therefore, the emulation

Table 9.2 The architecture of the operation emulation system of the container terminal

Layer	Contents
Presentation layer	UI presentation and data statistics API
Application layer	System business process
Logic layer	Business logic, environment constraints
Control layer	Intelligent planning, intelligent scheduling
Data layer	Emulation TOS, intelligent knowledge base of each subsystem

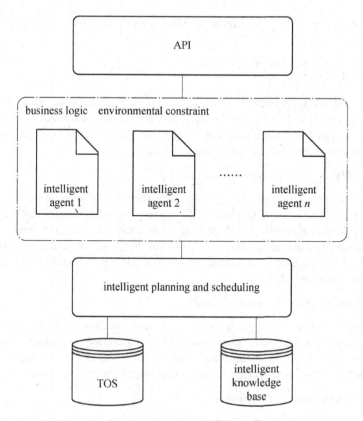

Fig. 9.17 Structure of the emulation system of the container terminal

system composed of this hierarchical multi-agent structure can be closer to the actual operation of the container terminal system and obtain better emulation effect.

9.4.2 Case of Ship Loading Emulation of the Container Terminal

According to the previous description of the architecture of the operation emulation system of the container terminal, taking the ship loading emulation as an example, the ship loading emulation process of the container terminal is mainly composed of TOS, various databases/data warehouses, intelligent planning and scheduling, as well as the agents and the operating environmental constraints/business logic of each agent. Details such as data transmission and data storage will be ignored in this section, and the ship loading emulation will be described from the aspects of intelligent planning and scheduling as well as intelligent scheduling and constraints of various agents involved.

1.   Intelligent planning and scheduling

(1)   Intelligent stowage plan

The stowage plan is determined according to the pre-arranged baplie provided by the shipping company and the current situation of container collection and storage in the container terminal. The intelligent stowage plan determines the stowage positions of the export containers that currently need to be loaded.

(2)   Intelligent loading instruction scheduling

According to the current facility operation and stacking situation of the container terminal, the instruction activation of the current period is determined with the stowage positions determined by the stowage plan as the constraint. Horizontal transportation facilities including the yard cranes and the container trucks are activated to complete the loading operation, according to the instruction and the subsequent scheduling and controlling.

(3)   Intelligent container truck scheduling

According to the activation status of the current instructions and container truck conditions, the container trucks are assigned to the activated instructions that need to be processed. After the assignment is completed, container trucks will begin to communicate with the yard crane or quay crane to continue the next step.

(4)   Intelligent yard crane scheduling

According to the distribution of current yard operation instructions and the position and operation situation of current yard crane, the activated instructions to be operated are assigned to specific yard crane, so that the yard crane (agent) can further schedule and decompose the tasks into yard crane operation instructions and fulfil the corresponding instructions.

2.   Environmental constraints and micro-scheduling of agents

The agents involved in the ship loading emulation of the container terminal include yard crane agent, container truck agent and quay crane agent.

(1)   Yard crane agent

According to the specific operation tasks assigned by the intelligent scheduling of yard crane, the yard crane agent is capable of further scheduling and decomposing the task instructions with its control system, and then generating the actual yard crane action instructions, and carrying out the execution control of these yard crane actions. For certain instructed actions that need to be coordinated with the container truck agent, it communicates with the container truck agent to complete the yard crane operation task, and gives feedback to the intelligent yard crane scheduling and emulation TOS. In this section, emulation in manual terminal is taken as an example. Considering the situation of automated container terminal, since some yard crane tasks require collaborative scheduling and control of multiple yard cranes, corresponding collaborative control modules should be added among yard crane agents for centralized collaborative scheduling to simulate part of collaborative control functions of ECS system in actual operation.

The yard crane agent needs to consider the following environmental constraints during the ship loading process.

a. Constraint on safe operating distance. When several yard crane agents operate simultaneously in the same block, the distance between the yard crane agents should be greater than the safe distance to ensure the safety of the yard cranes and the container trucks.

b. Crossing constraint. When the yard crane agents operate in the same block, they cannot cross each other.

c. Constraint on the maximum number of yard cranes in the block. Too many yard cranes cannot enter at the same time in the same block, so as not to cause the operation congestion of yard cranes and container trucks in the block. Multiple yard cranes running at the same time will cause excessive load on the track or roadbed, affecting the operation safety.

d. Constraint on the spreader height of the yard crane. When there are containers stacked higher between the container location to be operated and the container truck, the spreader of the yard crane should be lifted vertically to the safety height above the higher container, and then move horizontally to prevent the containers from scraping each other in the operation process, which may cause accidents in the operation.

The yard crane agent involves the following intelligent scheduling and controlling in the ship loading process.

a. Scheduling priority of yard crane tasks. According to the scheduling and allocation of yard crane operation tasks by the intelligent scheduling of yard crane, the yard crane executes secondary scheduling for all operation instructions to be completed by itself, based on the current condition of the container storage area and the condition of other tasks to be completed, which simulates the secondary scheduling of the task assigned to the driver of the yard crane in the actual operation of manual terminal as well as the decomposition and secondary scheduling of the yard crane task by the ECS system in the automated container terminal. After the completion of the scheduling, the order table of the current yard crane tasks to be executed will be generated, and then the tasks will be decomposed and executed according to the task priority order.

b. Action decomposition of yard crane. The operation tasks of the yard crane can be decomposed into specific action instructions. For example, if the current location of the yard crane is 1A020, the task of loading the container at location 1A0300503 to container truck T03 should be decomposed into the followings. Translating the gantry of the yard crane to 1A030 → translating the trolley of the yard crane to row 05 → catching the container at 1A0300503 with the spreader of the yard crane → communicating with the container truck

T03 → lifting the spreader and translating the trolley to the operation lane of the container truck after successful communication → lowering the spreader and landing the container onto the container truck → communicating with the container truck, confirming the landing of the container and lifting the spreader.

(2)   container truck agent

Based on the current tasks of the container truck to execute, the container truck agent analyzes and makes decisions of the travelling route, carries out processing such as avoiding in the process of driving, arrives at the destination according to safety rules, then communicates with yard crane agent and quay crane agent to complete current container truck task, and feeds back to the intelligent scheduling of container trucks as well as emulation TOS.

The following environmental constraints need to be taken into account in the ship loading process for the container truck agent.

a.   Constraint on collision avoidance. The container truck agents should avoid collisions when running in the same route.

b.   Constraint on avoiding deadlock. The container truck agents should avoid route deadlocks when running within the same block.

c.   Constraint on safe distance. When running in the same route, it should be ensured that the distance between the container truck agents is greater than the safe distance, ensuring the braking distance under special circumstances.

d.   Constraint on the maximum number of container trucks within the block. In order to ensure the smooth traffic and operation safety in the container storage area and other areas, the maximum number of container trucks in each block is usually limited to control the regional traffic flow. Therefore, the maximum number of container trucks in the block should be observed by the container truck agent in the emulation process.

e.   Crossing constraint. Container trucks are not allowed to cross each other in the same lane.

The container truck agent involves the following intelligent scheduling and controlling in the ship loading process.

a.   Route controlling. According to the ship loading instructions given by the container truck scheduling, the container truck agent decides the route to the destination location according to the current traffic flow conditions in the terminal, and updates the route at a specific node.

b.   Detection of deadlock in route. It is calculated in time that whether there is deadlock conflict in the route with other container truck agents when the route changes. If there is deadlock, the route is re-calculated.

c.   Driving controlling. In the same route, when multiple container truck agents drive in the same direction to the approximate destination, the speed of traffic flow is controlled, and the driving speed is controlled under the condition of ensuring the safe distance between container trucks, so as to improve the efficiency of synchronous driving.

(3)　Quay crane agent

According to the operation order requirements of the given operation line plan and stowage plan, the quay crane agent communicates with the container truck agents and completes the last step of ship loading under the constraints of safe handling, then conducts tally confirmation process after the container is loaded to the position in the ship, and feeds back to the scheduling system of ship loading instructions and emulation TOS.

The quay crane agent needs to consider the following environmental constraints during the ship loading process.

a.　Constraint on operation line order. In general, order of operation lines shall be ensured. For example, at the same operating bay, the operation shall be carried out in the order of 20-foot-single-lifting prior to 20-foot-double-lifting prior to 40-foot-single-lifting.

b.　Constraint on the order of container positions in the loading ship. In the process of ship loading, the order of loading tasks on the quay crane shall meet the requirements of the order of ship loading operations. For example, when loading in the cabin, it should be ensured that the containers located at the lower of the same row are loaded before the containers at the upper, and the loading position remains unchanged. In deck loading, in addition to ensuring the loading order of containers in the same row relative to the upper and lower positions, the "stair" operation should also be carried out from the sea side to the land side to ensure that no cross-container operation is carried out during the loading, so as to ensure the safety of the operation.

c.　Constraints on ship roll and torque. During ship loading operations, the ship roll and torque should be kept within a safe range in the operation order of the quay crane. When the safety critical point is reached, the loading order should be changed to adjust the center of gravity and torque.

Quay crane agent involves the following intelligent scheduling and controlling in the ship loading process.

Ship loading order rearrangement. According to the container trucks that have reached the quay crane communication area, the ship loading order shall be rearranged. The containers that can meet the requirements of the current shipping order shall be placed in the priority order for loading. If no containers can be loaded, waiting is necessary.

9.4.3　Optimization of Emulation

Since the emulation can simulate the internal structure and operation mechanism of the system and deduce the micro functions and capabilities of the system from the actual activity process of each substructure, the emulation can be used as an evaluation tool to estimate the specific cost and efficiency of the system in the micro scenario.

However, when the purposes of evaluation are different, the corresponding accuracy and efficiency requirements are different according to the purpose of evaluation. For example, when evaluating and comparing the decision results as a cost-efficiency evaluation tool, the evaluation tool is usually required to have higher accuracy so as to evaluate the decision results correctly. As an environmental scenario evaluation tool in AI and machine learning, evaluation tool is usually required to ensure high computational efficiency under the condition that bias is within the allowable range to ensure the training speed. Therefore, in order to solve the efficiency and effect problems of emulation under different evaluation purposes, the sub-model of emulation with different granularity can be used to achieve the balance between emulation efficiency and emulation accuracy to different degrees.

When establishing the emulation model, each sub-model is stratified by the detail degree of internal structure simulation, which is divided into coarse-grained sub-model, medium-grained sub-model and fine-grained sub-model. The coarse-grained sub-model preserves only the basic structure of the subsystem. The microstructure is simulated with major constraints considered. Bias correction is used to ensure that the sub-model is unbiased. Based on the basic structure of the sub-system, some sub-system structures that have little impact on the efficiency are added to the medium-grained model. The microstructure is simulated considering most of the constraints of the sub-system. Meanwhile, bias correction is used to ensure that the sub-model is unbiased. The fine-grained sub-model, on the basis of retaining all the microstructure of the subsystem, takes into account all the constraints of the subsystem for simulation, so as to ensure that the functions and capabilities of the system can be presented as completely as possible from the microscopic perspective.

Bibliography

1. Banks J, Carson J, Nelson B et al (2000) Discrete-event system simulation, 3rd edn. Pearson, London
2. Qi H, Wang X (2004) System modelling and simulating. Tsinghua University Press, Beijing
3. Zhang X (2005) Digital simulation of control system and CAD. China Machine Press, Beijing
4. Zhao N, Xia M, Mi C et al (2015) Simulation-based optimization for storage allocation problem of outbound containers in automated container terminals. Math Probl Eng 2015
5. Yang X, Mi W, Li X et al (2015) A simulation study on the design of a novel automated container terminal. IEEE Trans Intell Transp Syst 16(5):2889–2899
6. Ucar I, Smeets B et al (2017) Simmer: discrete-event simulation for R. J Statist Softw 90(2)
7. Yang X, Zhang W, Mi W (2019) Research on the method of storage planning for the large-scale container yard. In: International conference on advances in construction machinery and vehicle engineering (ICACMVE). IEEE, pp 255–260

Chapter 10
Smart Port and Digital Monitoring and Diagnosis

In the history of business evolution, the fundamental driving force for development is technological innovation. The most basic and important innovation is the so-called "general technology" or "meta-technology", which can open up a new era. After the emergence of these technologies, a large number of "complementary innovations" can continue to emerge on the basis of these technologies, further enriching and iterating the connotation of the business era.

For example, the steam engine, electric power, and internal combustion engine technologies of the first and second industrial revolutions are "general technologies". The emergence of these technologies made the previously scattered and unsystematic productive forces quickly gather together under the attraction of "scale effect". It was only after complementary innovations (such as automobiles, trucks, chainsaws, lawnmowers, large retailers, shopping malls, cross-docking warehouses, new supply chains, etc.) emerged that the industry's outline became clear. Just like the original scattered dots were connected, forming horizontal and vertical lines (industry).

The emerging digital technology is the next "general technology". The digital technology has broken down barriers between industries and will connect different social elements in an unprecedented way, thus opening up ecological space of higher order. Digital technology has created the connectivity of data, the connectivity of scenes, and the interoperability of values. These factors have created a new force that is stirring the once-well-organized divisions of the industry, and the boundaries of the industry are beginning to fall apart. Digital monitoring and diagnosis cover sensor network, IOT, AI, industrial monitoring, mobile edge computing (MEC), cloud computing construction and other core contents.

© Shanghai Scientific and Technical Publishers and Springer Nature Singapore Pte Ltd. 2022
W. Mi and Y. Liu, *Smart Ports*, https://doi.org/10.1007/978-981-16-9889-7_10

10.1   Overview of Digital Monitoring and Diagnosis

10.1.1   Concept of Equipment Condition Monitoring

In order to ensure the safe use of port machinery, suitable working conditions and operating conditions should be maintained, and various condition inspections should be carried out. Inspections are usually divided into routine inspection, regular inspection, periodic inspection, enhanced periodic inspection, special inspection and extensive inspection. Special inspections and extensive inspections are usually carried out by the third-party, professional institutions and engineers, while the rest are organized and carried out by the port. Personnel requirements, inspection contents and grades, and records of inspection results are provided in *ISO 9927–1:2013/GB/T 23724.1–2016*. Such inspections are usually considered as subjective condition inspections. However, due to the differences in port scale, professional ability of engineers and equipment management concept, there are many differences in the implementation and effect of port equipment inspection in different terminals.

It is called condition monitoring to check and identify the physical conditions of the whole machine and components of port logistics facility in operation, so as to determine whether the operation is normal or not, whether there are signs of abnormality and deterioration, or to track the abnormal situation, predict its deterioration tendency, determine its deterioration and wear degree, etc. The purpose of condition monitoring is to understand the abnormal performance and deterioration tendency of equipment before failure, so as to take targeted measures in advance to control and prevent malignant failure, thus reducing equipment downtime and loss, cutting equipment maintenance costs and improving equipment utilization rate. Equipment condition monitoring is the concrete implementation of fault diagnosis technology and the inspection technology to understand the dynamic characteristics of equipment, which includes the main non-destructive inspection technology, such as vibration monitoring, noise monitoring, corrosion monitoring, stress monitoring, temperature monitoring, leakage monitoring, wear particle analysis (ferrography), spectral analysis and other various monitoring technologies of physical quantities. Such monitoring is usually identified as objective condition monitoring.

On-line monitoring of port logistics facilities in operating condition without shutdown, is able to understand the actual characteristics of the facilities and help to determine the components that need to be repaired or replaced, fully explore and utilize the potential of the facility components and the whole machines, avoid excess and repeated maintenance, cut maintenance costs, reduce the production loss caused by shutdown. Especially for the key facilities that require efficient, continuous and stable operation in smart port, its significance is more prominent.

10.1.2 Concept of Digital Monitoring and Diagnosis

Digital monitoring and diagnosis is performed by measuring some relatively single characteristic parameters of equipment (such as current, voltage, temperature, pressure, vibration, stress, etc.), and processing, analyzing and extracting features of the measured signals combined with historical conditions, analyzing the data and making automatic calibration diagnosis according to the relationship between the values of characteristic parameters and threshold values, so as to quantitatively identify the operating conditions of mechanical equipment and its components (normal, abnormal and failed), and further determine what necessary measures need to be taken to ensure that the mechanical equipment is maintained in the optimal performance. Therefore, all the necessary measures are implemented with the hardware and software configured in the operation site of the equipment, which belongs to the parallel application of multiple single-physical domains with sensor networks. In the field of local data processing, edge computing or fog computing is often applied, which can respond within milliseconds, also known as a local AI solution.

The application of digital monitoring and diagnosis of higher level is to compare all the on-site operating systems with cloud computing, draw conclusions from changes in individual system to other systems via big data analysis, deep learning and optimization of AI algorithms, and expand application scenarios based on terminal equipment management, so as to achieve group AI of multi-physical domain fusion, namely, to provide reliability predictions for failures or alert changes in equipment conditions by algorithms. The structure of digital monitoring and diagnosis of smart port is shown in Fig. 10.1.

10.2 Development of Digital Monitoring and Diagnosis

At present, the codes and standards for digital monitoring and diagnosis in the port field are fragmented, and there is a lack of codes and standards considering the global digital informatization in the whole life cycle. For instance, the specific requirements of vibration signal processing are given in *Condition Monitoring and Diagnosis of Machines—Vibration Condition Monitoring—Part 2: Processing, Analysis and Presentation of Vibration Data (GB/T 19873.2–2009/ISO 13373–2:2005)*. In *Lifting Appliances—Safety Monitoring System (GB/T 28264–2017)*, only the standard requirements for crane safety monitoring and a simple and general description on the terms related to communication protocol are put forward. Other relevant codes and standards have similar problems.

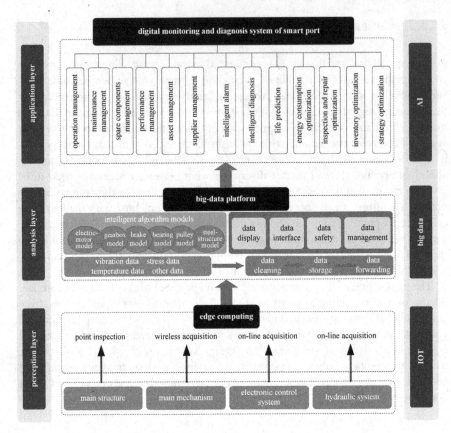

Fig. 10.1 The structure of digital monitoring and diagnosis of smart port

10.2.1 Basic Conditions of Digital Monitoring and Diagnosis of Smart Port

Port facilities are mainly composed of structures, mechanisms, electrical systems, hydraulic systems, winding systems and other systems. The electrical signal monitoring technology of the electrical system has been relatively mature. However, the monitoring data are mainly used to inquiry alarm information and support trouble shooting, and there is a lack of effective analysis and utilization of these data. The mechanism monitoring mainly uses vibration, temperature and other quantities to carry out routine monitoring, which can play a certain role in helping the reliable operation of port cranes and the continuity of port operation. The structure monitoring, which is crucial to port cranes, mainly relies on the stress monitoring at present. However, in the design of port machinery, extremely special conditions such as earthquake and storm are taken into consideration comprehensively, so it is difficult for the threshold setting of stress monitoring and the data calculated by design to effectively coordinate in analysis under the daily operating conditions, which also results in the greatly reduced monitoring effect. Limited by available sensor types, installation positions and cost restrictions, it is still difficult to have effective solution for vicious damage of crane such as crack damage, fatigue failure, slow and long-term structure deformation. Traditional monitoring plans are not competent for the diagnosis of such unsteady, nonlinear and non-classical conditions, not to mention the accurate prediction of failures in extremely complex conditions.

Port logistics facilities are dangerous, specialized, heavy, big, with long service period and of other properties. It is necessary to conduct system plan from a global perspective with digital thinking and implement digital monitoring and diagnosis at various stages within the whole life cycle from planning, design, manufacturing, purchasing, installation, operation, maintenance, modification, update, until final disposal. Due to the low degree of digitalization of the design source of port facilities, most of the design drawings are modified manually by traditional 2D CAD, and the manual input is carried out by finite element software for trafficability calculation. The design means are faced with the strong demand of digital upgrading. The production and inspection processes in manufacturing are mainly accomplished manually relying on a large number of labors. The BOM reports of all kinds of design, manufacturing and inspection are still in the form of documents and forms. Valuable design calculations and manufacturing data could not support the subsequent service of the facilities. The hidden troubles of facility conditions caused in many manufacturing processes are also transferred to the follow-up service cycle of the facilities, and the technical transformation and optimization of the port are difficult to be fed back to the design source of the facilities.

Operation reliability is an important parameter to measure whether the facility is operating normally, which is directly related to the economy and safety of the whole life cycle of port logistics facility. Reliability analysis and facility condition feedback are mainly used to evaluate and predict the reliability of facility during operation with operation data, and guide the design and optimization of the whole life cycle of facility, operation and maintenance engineering analysis based on the feedback analysis results. The transformation of facility design model into operation and maintenance model, prediction model requires a large amount of performance and environmental historical data to predict the accuracy of facility. It also takes considerable time to analyze field data to establish a normal performance baseline. Machine learning is used to determine the most efficient data steps in the data model to improve the data science automation.

With the rapid development and cost reduction of computing power and transmission bandwidth in various application scenarios, model-based (MBD) design, integration of computing and manufacturing, integration of process and manufacturing, 5G communication, digital twinning, machine learning and other technologies have rapidly become practical. Port logistics facility operation is a typical case of non-linear and non-modeled operation. Digital twinning, machine learning, IIOT and other new technologies will become the main tools in digital diagnosis. With the continuous introduction of various emerging technologies in the port field, conditions will be provided for smart port logistics facilities to realize the operation and maintenance scenario of self-perception, self-adaptation, self-learning, self-evaluation and self-decision-making.

10.2.2 Visualization of Equipment Monitoring and Diagnosis

Visualization of equipment management platform in full life cycle, including equipment modeling visualization, equipment manufacturing process visualization, equipment installation management visualization, equipment operating condition visualization, equipment ledger management visualization, equipment inspection management visualization, etc., is embodied in geometric modeling of equipment, which can intuitively, truly and accurately display the equipment shape, equipment distribution and equipment operating condition. Meanwhile, it can bind the equipment model with reality, archive and other basic data to realize the rapid positioning and basic information query of the equipment in the 3D scene.

(1) Visualization of equipment modeling refers to the 3D modeling of equipment components, equipment assemblies and the whole machine by using 3D modeling technology, establishment of 3D model library of components and equipment, and the display of hierarchical relations among the whole machine, components and parts, so as to realize interaction between people and 3D objects in the scene.

(2) Visualization of equipment installation management refers to the 3D modeling of equipment installation, combining 3D scenes with planning and actual schedule time, and displaying the installation and construction process of each stage with different colors.

(3) Visualization of equipment operating condition refers to the whole process of equipment operation management, including the 3D modeling of equipment and the operation condition visualization based on digital twin of on-site equipment, as well as the historical failure, current condition, operation trend analysis and planned maintenance of equipment.

(4) Visualization of equipment ledger management refers to by establishing equipment ledger and asset database, and binding with 3D equipment, to realize the visualization of equipment ledger as well as mutual search of model and attribute data and bidirectional retrieval of location, so as to achieve 3D visualization of asset management, enabling users to quickly find the corresponding equipment, and view the equipment actual situation such as the corresponding site location, environment, associated equipment, equipment parameters, etc.

(5) Visualization of inspection management means that all inspection tasks from formulation, allocation, issuing, reception, execution and assessment can be controlled remotely and synchronized real time with wireless, so as to realize the visualized, simplified, standardized and intelligent management of the inspection process, so that users can find equipment defects and various security risks in time.

10.3 Applications of Digital Monitoring and Diagnosis in Smart Port

From the end of the on-site sensor network, to the tube of data transmission, and the cloud of the comprehensive analysis of big data, there is a lot of engineering work to be done at all levels. The digital monitoring and diagnosis of equipment is a developing process combining the development of new technology with the manual daily inspection, maintenance, and equipment fault diagnosis for a long time.

10.3.1 Real-Time on-line Intelligent Condition Monitoring and Fault Analysis System for Reducer

Reducer is a key component of lifting facilities in terminal. In the actual operation of terminal, there will be some mechanical failures related to reducer. These reducers are usually installed in the machine rooms tens of meters above the ground. Once the failure cannot be found in time, it will result in continuous damage to the reducers, seriously affecting the economic benefits of the terminal. On-line monitoring and fault diagnosis of the reducer in the key positions can understand and evaluate the operating condition of the equipment in real time, and make the fault diagnosis and forecast in advance. By transforming the shutdown after failure into planned shutdown, reducing the shutdown time or avoiding the deterioration of the accident, and gradually transforming the planned maintenance and maintenance after accident into preventive maintenance, it can improve the modernization level of enterprise equipment management and create huge economic benefits.

The real-time on-line intelligent condition monitoring system for reducer is based on the directional parameters such as vibration and temperature. The system mainly includes four parts: front-end perception layer, data acquisition-analysis-diagnosis module, user database server or cloud server, Web server and browser. The overall architecture is shown in Fig. 10.2.

The front-end perception layer includes various sensors, such as vibration, bearing temperature, gear oil temperature and granularity, etc., which are usually installed in the reducer (Fig. 10.3). The data acquisition-analysis-diagnosis module is usually installed in the electrical room, which is used to realize the data acquisition, filtering, analysis and self-diagnosis of each sensor. Acquisition software and database are usually installed in central control room for data storage, precise analysis and big data management. The IOT is used as the data exchange medium between each functional module. The data acquisition and analysis module configures the perception layer devices and collects signals by reading the acquisition parameters from the database server, and transmits real-time data through the central control room servers such as 4G/5G/ Wi-Fi/Ethereum network. As the client, the central control room servers realize parameter configuration, signal analysis, condition monitoring, fault diagnosis and big data management. When data need to be viewed remotely,

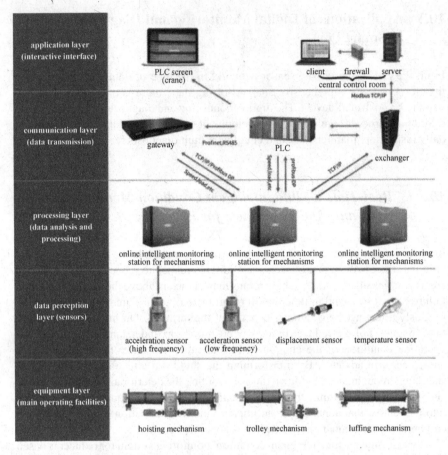

Fig. 10.2 The overall architecture

the browser makes a request to the Web server to access the database server or cloud server to view the historical data. The data analysis and diagnosis module is shown in Fig. 10.4.

The online monitoring systems usually monitor the following types of data.

(1) Vibration. The vibration signal contains abundant fault information, which can reflect the operating condition of the gears more quickly and intuitively.

(2) Bearing temperature. Bearing temperature is an important monitoring data of the reducer. When the reducer is operating, the bearing friction will produce a certain amount of heat. When the heat generated reaches thermal balance, the bearing temperature will be within the normal range. When the bearing is abnormal, the thermal balance will be destroyed and the bearing temperature will rise rapidly. If the bearing temperature exceeds the preset temperature, the reducer should be stopped immediately for check.

Fig. 10.3 Front-end
acquisition device

Fig. 10.4 The data analysis
and diagnosis module

(3) Gear oil temperature. When the reducer is operating normally, the oil temper-
    ature should not be higher than 90 °C. When the oil temperature is too high,
    the monitoring system should display warning.

(4) Metal wear particles in oil. Through real-time monitoring of the tiny wear
    particles of metal elements in gear oil, gear and bearing wear as well as gear
    oil life can by analyzed with big data analysis.

The real-time on-line intelligent condition monitoring and fault analysis system for reducer has the following technical advantages.

1.   Complete database resources

Database is the key part of on-line monitoring and fault analysis system. Equipment information, component information, sensor information, measuring point information, analysis results, alarm data and fault samples are stored in the database in a prescribed form. As the port crane leader, the system developer has the accumulation of complete design data and fault statistics data of port equipment. The powerful expert team of the system developer, can accurately grasp the reducer operating condition through the real-time operation data acquisition of the reducer, and use the powerful data cloud computing service platform, to analyze and evaluate the reducer by comparing and combining the operation data and the design data, so as to accurately evaluate the operation condition of the reducer.

2.   Intelligent online vibration monitoring system

The intelligent online vibration monitoring station can collect the reducer vibration data in real time, and implement data compression, filtering, differential processing and indication value calculation locally, then transmit the calculation results to PLC and central control room server to quickly respond to equipment failure. The intelligent online vibration monitoring system has the following characteristics. (1) Full sampling technology has been applied. (2) Full synchronous acquisition technology has been applied. (3) Condition acquisition and triggered storage technology has been applied. (4) The intelligent collector has good expansion characteristics and can access signals such as vibration, speed, liquid level and temperature. (5) Automatic and intelligent diagnosis function has been achieved.

3.   Edge algorithm and intelligent diagnosis

Compared with other systems, the intelligent collector developed by the system has the ability of edge computing, which integrates the functions of acquisition, adaptive filtering, FFT calculation, broadband and narrowband energy calculation, automatic diagnosis and alarm. The technicians of the system developer have programmed the equipment fault feature signals (such as bearings) and the analysis experience of professional engineers (reducer diagnosis) into the algorithm built inside the collector. The collector will automatically output the diagnosis results, thus reducing the dependence of the on-site service engineers on the specialization of the vibration signal analysis system.

10.3.2 TRUCONNECT Remote Monitoring for Crane

The intelligent remote monitoring system first developed by a company was used in the company's travelling cranes in a nuclear power plant, which could work under the extremely harsh conditions of the intelligent monitoring system. Subsequently, this system was applied to the company's industrial cranes, and catalyzed its wide application in the company's mobile port cranes, straddle carriers and other fields. So far, TRUCONNECT has been used on nearly 20,000 machines worldwide by the company.

In 1995, while still an engineer at SMV, Rogers designed the CAN bus system for heavy forklifts. In the next year, SMV combining with Parker, a hydraulic drive company, injected the important part of IQAN control system into the whole CAN bus system. The machine integrated intelligent detection function and safe operation protection function, which could flexibly deal with various difficulties and complications in operation. Meanwhile, the efficiency of hydraulic and transmission system was greatly improved. The whole machine was brought into the real intelligent era. In addition, IQAN reduced the average fuel consumption by more than 30% and greatly improved the durability of the entire machine components. This environment-friendly machine significantly improved the utilization of resources and also contributed to reducing the pressure on the earth's resources and the environment.

After joining the group, SMV successfully connected TRUCONNECT and IQAN systems into an efficient and collaborative intelligent network. Integrating multiple technologies, the company finally formed the most reliable remote monitoring and cloud service system in the container handling industry. Diversified and flexible TRUCONNECT could set up different sensors and data items according to the lifting needs of different users. In addition to the basic functions of monitoring, analysis and early warning, it has added some functions such as geo-location-based service and the weighing function embedded in the international SOLARS certificate system. The diversified and open system attributes have made TRUCONNECT possess unlimited potential in the future.

The TRUCONNECT remote monitoring system uses sensors to collect utilization data, including operation time, motor start, service cycle and braking condition. In the cases of crane overloading, emergency shutdown and excessively high temperature, the warning will be sent by text message or email. It also provides brake and inverter monitoring, predicting the remaining designed working period (DWP) of the selected components, such as cranes and crane brakes. With the condition monitoring device installed on the equipment, TRUCONNECT collects data, which are then sent to a remote data center for integration, so as to provide customers with round-the-clock connectivity to a global network of crane experts with the remote supporting system, assisting in problem solving and troubleshooting to reduce unplanned downtime. The interface of TRUCONNECT remote monitoring system is shown in Fig. 10.5.

The system supports any connected device across platforms, and the information is presented one by one in the form of easy-to-understand graphs and charts. The system displays TRUCONNECT operation data and warnings, and provides analysis

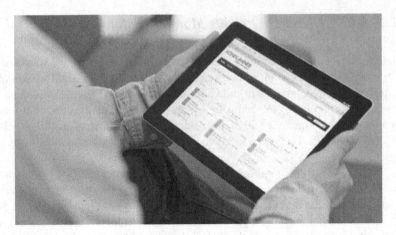

Fig. 10.5 The interface of TRUCONNECT remote monitoring system

of patterns and trends. For example, repeated overheating warnings mean equipment or processes may need to be adjusted. Researching trends can help distinguish the priorities of corrective measures and investment behaviors. Data and behavior analysis over time improves the feasibility of predictive maintenance. The home page of the supporting system's website will display the failure history and rank the equipment utilization based on selected criteria for quickly examining the pain points of the entire group of equipment. The system also provides data archiving and retrieval options, including file uploading and electronic reports for printing.

The company is currently exploring partnerships with Nokia in digital automation cloud technology, leveraging advanced networks and the global digital automation cloud technology to take advantage of future network technologies, and now is capable of developing 5G.

10.3.3   Application of on-line Automatic Wire Rope Inspection System on Quay Crane Based on Weak Magnetic Detection Principle

Wire ropes of port cranes are consumable products with friction loss. As key components, their fracture accidents are important hidden troubles that have not been solved for a long time in port industry. Due to the complex working conditions and the long-term use under heavy load and high speed, damage such as broken wires, wear, broken strands, rust, fatigue, is very easy to happen. If there is no efficient and reliable safety monitoring means, the defects of the wire rope cannot be found in time and management measures not be taken, which may easily lead to the malignant accident of the wire rope fracture. At present, there are 4 difficulties for the effective inspection of traditional wire rope.

1. The wire rope is difficult to effectively inspect due to its various states. According to the State Administration of Work Safety survey data from over 8000 wire rope users, more than 10% of the wire ropes in use have lost more than 15% of their strength, which are in dangerous condition; more than 2% of wire ropes in use have lost more than 30% of their strength, which are in extremely dangerous condition. Because the wire ropes are always in a state of wear and tear, there are all kinds of "causes" for strength loss and the risk of fracture accidents. In the field of port logistics equipment, casualties and property losses caused by wire rope fracture occur every year.

2. It is inefficient to check the wire rope manually. According to the provisions in *Wire Rope Spot Inspection and Update System* for port enterprises, for the wire rope in the first 18 months after putting into operation, inspection shall be conducted once a month; from the 19th month to the 22nd month, the inspection shall be conducted once every half month; from the 23rd to the 24th month, the inspection shall be conducted once a week. Each inspection needs 5 people and takes 4 h. The quay crane needs to be shut down during each inspection. Within the life cycle of 2 years of the most important lifting wire ropes, the time of manual inspection for shutdown is about 136 h. In addition, the quay crane driver needs to visually check the wire ropes for about 10 min every day. The manual inspection mainly relies on visual and simple tools, which not only wastes a lot of time in operation, but also results in unsatisfactory inspection effect.

3. The manual inspection results of wire ropes are unreliable. In the manual inspection, limited by the sight of the eyes, only the part visible to the naked eyes can be observed, while the opposite part of the wire rope cannot be seen. In addition, due to the influence of the environment, light and other factors, in many cases, visual measurement is basically a skim, the inspection effect is difficult to ensure. In some locations, the inspectors can only be in the place with the aisle railing, which is very far away, or the viewing angle is limited. Even in full accordance with the speed of 0.3–0.5 m/s for inspection, manual inspection can only detect obvious external defects of wire ropes, such as broken strands, while the internal wear, rust and broken wires, especially fatigue and other damage are more difficult to detect. According to previous studies, when the external wires of the wire rope began to break, the undetected broken wires inside are generally 2.5 times in amount of that can be seen outside.

4. The cost of wire rope is high and waste exists. Due to the limitation and uncertainty of manual inspection, in order to ensure the safe use of the wire ropes and avoid major accidents, the port enterprise can only take regular replacement of the wire ropes with fixed container handling volume. No matter how the actual condition is, the wire ropes should be replaced as long as the specified time or the specified container volume is reached. Some port enterprises make provisions on the wire rope replacement of quay cranes. Under the condition of regular maintenance and lubrication every quarter, for the imported wire ropes, the hoisting wire ropes are replaced every 2 years, the trolley traction wire ropes are replaced every 2.5 years, and the pitching wire ropes are replaced every

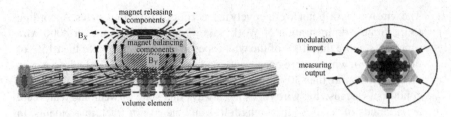

Fig. 10.6 The principle of weak magnetic detection for wire rope on-line monitoring

8 years. According to the foreign research, about 70% of replaced wire ropes have little or no strength loss, which causes a lot of waste and the unnecessary downtime.

The on-line automatic inspection system for wire ropes, integrating magnetic memory programming method, weak magnetic sensor technology, pattern recognition technology, modern network communication technology, etc., can efficiently and accurately online detect broken wires, wear, rusting, fatigue, deformation and other kinds of damage of wire ropes, so as to find hidden dangers in the bud, and avoid safety accidents. Moreover, the online automatic inspection system for wire ropes can carry out online daily evaluation of the wire rope damage, regular evaluation of the damage development trend, real-time warning of the sudden major hidden danger, etc., and solve the technical problems that have troubled the nondestructive inspection of quay crane wire ropes for a long time. The principle of weak magnetic detection for wire rope on-line monitoring is shown in Fig. 10.6.

The on-line inspection system for wire ropes strictly follows the principle of stress checking for wire ropes stipulated in the international standard, combining with a large number of practical data of manual nondestructive detection, develops and constructs professional processing software, establishes the core algorithm in line with the fact, to conduct the predictive research on the development condition of wire ropes. The damage of wire rope is divided into 5 grades by professional definition. The damage condition of the loss rate ranging from 2.5 to 9.5% of the stressed section of wire ropes can be detected online throughout the whole period, and the detection rate is 83–100%. The inspection range of the hoisting wire ropes covers 95% of the working sections. Moreover, the defect positions of the wire ropes can be accurately located at high speed. The detection results can be sent to the central control room in real time through the Internet, and the real-time warnings can be made for the detected results beyond the standard, which can realize the intelligent synchronous monitoring at high speed in the whole life cycle of the wire ropes. The on-site monitoring of hoisting wire ropes, trolley traction wire ropes and pitching wire ropes is shown in Figs. 10.7, 10.8 and 10.9.

Online automatic monitoring system for wire ropes can realize online automatic inspection of quay crane wire ropes, forming a complete and traceable inspection report. Comparison results of online automatic monitoring and traditional manual inspection for wire ropes are shown in Table 10.1.

Fig. 10.7 The on-site monitoring of hoisting wire ropes

Fig. 10.8 The on-site
monitoring of trolley traction
wire ropes

Application of on-line automatic monitoring system for quay crane wire ropes, can greatly improve the security of quay crane wire ropes and intelligent level of inspection, realize the online machine inspection to replace manual regular inspection, while effectively put an end to hidden trouble of wire rope safety, reduce the labor intensity of workers, improve the service cycle of wire ropes, and reduce unnecessary downtime, which can be of great help to improve the quality and efficiency of port enterprises. According to the operation parameters of the quay crane steel wire ropes during operation and design parameters of the wire ropes before delivery, combining with the inspection parameters, maintenance information and historical

Fig. 10.9 The on-site monitoring of pitching wire ropes

Table 10.1 Comparison results of online automatic monitoring and traditional manual inspection for wire ropes

Comparison item	Online automatic monitoring	Manual inspection
Inspection method	Online automatic inspection, not affected by the operation speed of wire rope, and can be completed under the normal operation environment	Manual inspection with visual measurement, hand touch, caliper, operation speed of wire rope within 0.3–0.5 m/s
Inspection time	All year round, machine intelligent inspection replaces manual inspection	Each inspection requires 5 people and takes 4 h, or 136 h in 2 years
Energy consumption	The inspection is completed during operation, and the power-on energy used exclusively for the inspection is negligible	The motor involved in each manual inspection consumes 3832 kW h, and 130,000 kW h in 2 years
Inspection results	It can effectively and accurately detect various kinds of damage such as broken wires, wear, rusting, fatigue, deformation and so on inside and outside the wire ropes, and completely eliminate the hidden danger of fracture accident	Manual inspection can only partially detect the external damage of the wire ropes, while the internal broken wires, wear and tear, rusting, especially fatigue and other conditions of the wire ropes cannot be detected
Inspection efficiency	The operation time of the hoisting wire ropes in two-year life cycle has been increased by 136 h. According to the single quay crane lifting 30 TEU per hour, the lifting container volume can be increased by 4080 TEU	During the inspection, the quay crane is shut down. Within the two-year life cycle of the hoisting wire ropes, the time of manual shutdown inspection is about 136 h

(continued)

Table 10.1 (continued)

Comparison item	Online automatic monitoring	Manual inspection
Replacement reference	Evaluates the wire ropes' using condition correctly and provides the scientific reference for the wire rope replacement	Replaces wire ropes regularly or quantitatively
Labor cost	Machine intelligence inspection replaces manual inspection, cutting a lot of labor costs	Each inspection of quay crane wire ropes needs 5 people, takes 4 h and consumes a lot of labors
Replacement cost	The safety condition of the wire ropes is always controllable, which provides the inspection reference for prolonging the service life of the wire ropes scientifically	Regular replacement of the wire rope causes a waste in cost
security	The whole life cycle of the wire ropes is safe and controllable, which fundamentally guarantees the use safety of the wire ropes	The internal broken wires, wear, rusting, fatigue are difficult to detect, so there are often major hidden dangers
Scientific management	All the inspection results can be automatically saved, and the inspection data can be traced at any time to improve the management level of the equipment scientifically	Records are scattered with poor traceability, and the life cycle of the wire ropes cannot be scientifically managed

data of wire ropes, the safety information of quay crane wire ropes is analyzed and provided, so as to provide the relevant personnel with real-time knowledge of the development stage of the life cycle of the quay crane wire ropes, provide high quality service for the continuous and reliable operation of the quay crane wire ropes, and provide strong guarantee for the normal operation of the quay cranes.

Bibliography

1. Ziqi X, Weijian M, Yifei S (1995) Synthetic diagnoses of large rotary support faults. J Shanghai Maritime Univ 4
2. Singh GK (2003) Experimental investigations on induction machine condition monitoring and fault diagnosis using digital signal processing techniques. Electric Power Syst Res 65(3):197–221
3. PaladeLakhmi V, Cosmin J, Bocaniala D (2006) Computational intelligence in fault diagnosis. Springer Science & Business Media
4. Liu Y, Mi W, Zheng H (2008) Structural fatigue assessment and management of large-scale port logistics equipments. Int J Comput Appl Technol 32(4):282–289
5. Isermann R (2011) Fault-diagnosis applications: model-based condition monitoring: actuators, drives, machinery, plants, sensors, and fault-tolerant systems. Springer Science & Business Media

6. Liu Y, Mi C (2013) Influence of crack size on the propagation trend of multiple rail surface cracks under rolling contact fatigue. J Appl Sci 13(19)
7. Wang J, Ye L, Gao RX et al (2019) Digital Twin for rotating machinery fault diagnosis in smart manufacturing. Int J Prod Res 57(12):3920–3934
8. Xu Y, Sun Y, Liu X et al (2019) A digital-twin-assisted fault diagnosis using deep transfer learning. IEEE Access 7:19990–19999

Chapter 11
Development Trend and Target of Smart Port

11.1 Development Trend of Main Hot Technologies

11.1.1 Development Trend of IOT

After 10 years of rapid development, especially the development of wireless network, cloud computing, big data technology and AI, IOT has deepened the whole process of social informatization and network. In particular, the application of the cyber-physical system has made many complex problems that could not be solved before get new solutions, providing unlimited vitality and space for the development of science and technology, giving birth to great changes in the society, and changing the traditional concept of security and benefit.

Industry 4.0 is defined in Germany as "intelligent production based on the cyber-physical system". Zhang Zhaozhong, a famous Chinese military theorist, also predicted that the Internet would enter the second half, and the new IOT would be iterated, which would be the core of the fourth industrial revolution. In his speech *The Fourth Industrial Revolution is Coming* on February 19th, 2020, he said that after this epidemic, some industries would emerge, popularize and upgrade rapidly, such as the IOT and smart logistics. Everything would have its own identity and be connected to the IOT via 5G. In the Internet era, man–machine interconnection is the solution to social problems, and the service industry has developed greatly. The IOT has solved the problem of physical interconnection between objects, and supported by 5G, it will fundamentally change our cognition, improve our lives and change the whole world.

In the future, the development of the IOT will interact and promote each other with the continuous development of a series of related industries, such as microelectronics technology, sensing technology, automatic control technology and AI.

First, for the sensors on the perception layer, they are the underlying components for sensing all kinds of information and data, and the basis and key for data and information collection in the IOT. At present, the high-end sensor market is mainly controlled by foreign suppliers, while domestic companies mainly focus on

© Shanghai Scientific and Technical Publishers and Springer Nature Singapore Pte Ltd. 2022
W. Mi and Y. Liu, *Smart Ports*, https://doi.org/10.1007/978-981-16-9889-7_11

the segmentation of sensors. In the future, there is a large space for domestic products to compete with foreign products in the segmentation of sensors. Meanwhile, with the development of the IOT industry, the demand for sensors is gradually increasing, the size of the sensor and power consumption also have a higher request, so the use of Micro-Electro-Mechanical System (MEMS) will be more popular, and it will gradually become the mainstream product of sensors in the IOT era. There are also non-contact image recognition technology and video recognition technology, which will be more and more applied to the perception layer.

Secondly, from the perspective of the network layer, with the development of 5G technology, 5G network will be more and more applied to the dynamic monitoring and supervision of objects in the IOT. In the future, the high-speed transmission rate with peak speed of 10 Gbps in 5G technology will effectively solve the problems of slow response and poor monitoring effect of existing video surveillance, and provide monitoring data faster with higher resolution. On the other hand, the property of multi-connection of 5G will promote safety monitoring to further expand its scope, getting monitoring data in more dimensions. It will provide more comprehensive and multi-dimensional reference data for the coordinated supervision of relevant national functional departments such as safety supervision, public security, fire protection, transportation and environmental protection, realizing the sharing of information, which is conducive to further analysis, decision-making and development of more effective safety measures. In addition to the field of video surveillance, 5G technology, with the help of AI, can not only realize intelligent perception, decision-making and early warning with data from various condition sensors, but also effectively reduce the excessive reliance on manpower and high cost in the traditional safety supervision field. Meanwhile, the most real-time, most vivid, most real data information in the field of safety supervision can also be obtained by intelligent means, and calculated accurately, so that all regulatory resource service deployment, regulatory manpower allocation and emergency response strategy will be more scientifically, accurately and effectively controlled. This will provide a strong guarantee to ensure that safety protection and disaster response are done at the right time, and that safety protection shifts from passive to active, from extensive to sophisticated.

Finally, from the perspective of the application layer of the IOT, the PaaS cloud platform of the IOT is the core of the industrial chain, on which all kinds of applications can be developed, deployed and operated. Therefore, the application and intelligence of the IOT are embodied on the cloud platform. However, cloud computing relies heavily on the network. Once the network is disabled, it will be impossible to obtain cloud computing. Therefore, for the research in the future, IOT will focus on the joint application of cloud computing and edge computing. The computing will be divided into two levels, that is, the part that only needs local data to achieve intelligent control will be placed on the edge computing layer, while the part that needs multi-party data fusion to achieve intelligent control will be placed in the cloud computing center.

In terms of application terminals of the IOT, the current applications of the IOT mainly include wireless payment, electronic consumption, Telematics, security monitoring and wireless gateways, etc. In the future, remote control, remote intelligent meter reading, transportation and public safety will also become important fields in infrastructure to realize the growth of IOT applications.

The research of AI will still be one of the hot issues in the future research of the IOT technology. On May 13th, 2018, the Cyber-Physical Systems Summit Forum and the First Cyber-Physical Systems Artificial Intelligence Conference was held in Beijing Convention Center. With the theme of "focusing on AI", the conference aimed to deeply implement the spirit of the 19th CPC National Congress, implement the deployment of "strengthening the R&D and application of new generation AI" at the 2018 NPC and CPPCC, and promote exchanges and cooperation between industry, education, research and application of AI in cyber-physical system.

Cybersecurity issues will also become a focus of attention in the IOT technologies. Starting from December 2019, Cybersecurity of Classified Protection 2.0 had been implemented, in which the IOT extension requirements would be applicable to all enterprises involved in the IOT equipment manufacturing, operation and industries closely related to such equipment, these enterprises must comply with the regulations of Cybersecurity of Classified Protection 2.0. On March 24, 2020, Aliyun announced that its IOT security platform (Link Security) had successfully passed the security assessment of the IOT based on Cybersecurity of Classified Protection 2.0 (Level 3), becoming the first IOT security service platform passing the assessment in China.

11.1.2 Development Trend of Blockchain

As the blockchain technology continues to explore and evolve, some new technological improvements have emerged. Blockchain, as a disruptive and innovative technology, has infiltrated into the fields of finance, assets, copyright, law and healthcare and become a new power of business growth.

1. Open source is becoming the main mode of public chain technology innovation

Open source has always played an important role in software technology innovation and contributed to the development of public chain innovation. The innovation mode of open source has made the public chain have the opportunity to gather the intellectual resources from all over the world and make them participate in the continuous development and optimization of the system, which has greatly reduced the workload of repeated innovation and improved the innovation efficiency. Open-source mode has almost been all adopted in the technology innovation of public chain. According to the CCID Global Public Chain Technology Assessment Project, 37 world-famous public chains as the assessment objects have all adopted open-source mode.

2. Data and scenarios have become important power of blockchain-driven innovation.

Blockchain is essentially an innovation in the way of data generation, calculation, sharing and storage under different scenarios. The scenario-based application of blockchain technology relies heavily on real scenarios and data. Emerging technologies and application innovations such as blockchain and AI have reached a new height in their desire for and dependence on data under various scenarios.

11.1.3 Development Trend of AI

1. Development from specialized intelligence to general intelligence

How to realize the leaping development from the specialized AI to the general AI is not only the inevitable trend of the next generation of AI development, but also a major challenge in the field of research and application. In October 2016, the National Science and Technology Council of the United States released the National Strategic Plan for Artificial Intelligence Research and Development, which proposed that the United States should focus on general AI research in its medium and long-term development strategy of AI.

2. Development from AI to human–machine hybrid intelligence

Learning from brain science and cognitive science is an important research direction of AI. Human–machine hybrid intelligence aims to introduce human function or cognitive model into AI system, to improve the performance of AI system, make AI become a natural extension of human intelligence, and solve complex problems more efficiently with human–machine cooperation. In China's new generation of AI planning and the American BRAIN Initiative, human–machine hybrid intelligence is an important research direction.

3. Development from "artificial plus intelligence" to the autonomous intelligent systems

Many of the current research methods in the field of AI rely more or less on artificial experience. For example, it is very time-consuming and laborious to manually design deep neural network model, manually setting application scenario, manually collecting and annotating a large amount of training data, and manually adapting the intelligent system required by user. However, a series of researches on autonomous intelligence also emerged gradually. For example, AlphaGo Zero, without the experience and knowledge of Go experts, mastered Go by playing on its own, and even found some strategies that had never been discovered by Go experts. Therefore, autonomous intelligence will be the research focus of the next stage of AI.

4. Inter-infiltration of AI with other fields of science

AI is a comprehensive frontier discipline and a highly intersecting interdisciplinary subject, with a wide and extremely complex research scope, and its development needs to be deeply integrated with computer science, mathematics, cognitive science, neuroscience and social science. With breakthroughs in super-resolution optical imaging, optogenetic regulation, transparent brain and somatic cell cloning, a new era of brain and cognitive science has been ushered in, with the ability to analyze the basis and mechanisms of neural circuits of intelligence on a large scale and in greater details. AI will enter a phase of biologically inspired intelligence, relying on discoveries from disciplines such as biology, brain science, life science and psychology to turn mechanisms into computable models. Meanwhile, AI will also promote the development of brain science, cognitive science, life science and even traditional sciences such as chemistry, physics and astronomy.

5. Booming of AI industry

With the further maturity of AI technology and the increasing investment from the government and industry, the cloud application of AI will continue to accelerate, and the scale of global AI industry will enter a period of rapid growth in the next decade. In 2016, an Accenture research report pointed out that the application of AI technology would inject new impetus into economic development, which could increase labor productivity by 40% on the current basis. In 2018, a McKinsey study predicted that by 2030, about 70% of companies would adopt at least one form of AI, adding $13 trillion to the AI economy.

6. Promoting mankind to enter the generalized intelligent society

The innovation mode of "AI plus X" will mature with the development of technology and industry, having a revolutionary impact on productivity and industrial structure, and promoting mankind to enter a generalized intelligent society. In 2017, the International Data Corporation (IDC) in the AI white paper, pointed out that AI would boost industry operation efficiency in the next five years. The China's economic and social transformation and upgrading have a high demand on AI. Under the demand of consumption scenario and industrial applications, it is necessary to break the perception bottleneck, interaction bottleneck and decision-making bottleneck of AI, promote the integration and improvement of AI technology and all walks of life in society, build several benchmarking application scenario innovations, and realize a low-cost, high-benefit and wide-ranging generalized intelligent society.

7. Increasingly fierce international competition in AI

At present, the international competition in the field of AI has begun, and will become increasingly fierce. In April 2018, The European Commission planned to invest $24 billion in AI from 2018 to 2020. The French AI Strategy, announced by the French President in May 2018, aimed to usher in a new era of AI development and make France an AI power. In June 2018, Japan released the Future Investment Strategy 2018, which focused on promoting the construction of the IOT and the application of AI. The world's military powers have also gradually formed a competitive situation with the acceleration of the development of intelligent weapons and equipment as the core. For example, the first National Defense Strategy report released by the US government sought to maintain military advantages through technological innovations such as AI to ensure that the US could win future wars. Russia proposed in 2017 that the military industry would embrace intelligence to strengthen the power of traditional weapons such as missiles and UAVs.

8.   The sociology of AI on the agenda

In order to ensure the healthy and sustainable development of AI and make its development achievements benefit the people, it is necessary to systematically and comprehensively study the impact of AI on human society from the perspective of sociology, and formulate and perfect AI laws and regulations, to avoid possible risks. In September 2017, the United Nations Crime and Justice Research Institute (UNICRI) decided to set up the first UN Centre for Artificial Intelligence and Robotics in The Hague to regulate the development of AI.

11.2   Development Trend of Smart Port

The construction and development of smart port is an important measure to implement the *Program of Building National Strength in Transportation* and accelerate the construction of world-class, green, smart and hub ports. It is also a powerful starting point to serve the overall national economy and jointly build "One Belt, One Road". It is also the only way to accelerate the transformation and upgrading of ports and improve the quality and efficiency of port enterprises. The premise and foundation for the development of smart port lie in the perfect integration of port functions and high and new technologies such as the IOT, industrial Internet, big data, cloud computing, 5G, blockchain, AI, etc. In practice, the main way for the development of smart port is the application and promotion of automated terminals and the intelligent reform and transformation of traditional terminals. The wide, systematic and deep applications of the comprehensive intellisense with the architecture of the IOT, knowledge, experience, information gathering based on the big data, intelligent decision-making and intelligent control technology with AI as the core, have built the main ecological system of the port construction and development, and embodying the main trend of smart port construction and development.

In the development process of this trend, the construction and operation of automated terminals and the intelligent reform and transformation of traditional container terminals have changed the most basic business forms of ports. Based on intelligent technology, the process of handling operation and the operation function of planning and scheduling are organically integrated. Figure 11.1 shows the change of integration of this function of the container terminal.

The new changes and demands of the basic functions of ports are highlighted in the following aspects.

(1)   Mechanical facilities are transformed from stand-alone control to system collaborative control, and equipment is transformed from automatic to intelligent.

(2)   Operation organization is transformed from master–slave centralized control of planning, scheduling and implementation to distributed cooperative control mode.

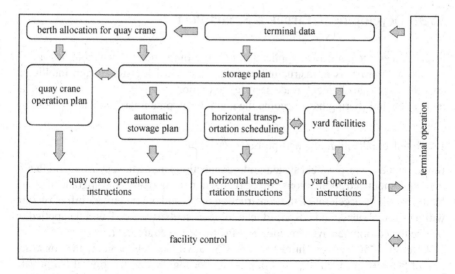

Fig. 11.1 Integrated control of terminal operations

(3) The big data analysis and intelligent operation management of port logistics system are organically combined.

(4) Intelligent coordination of human, machine and environment is matched with efficient and safe high-quality services.

(5) Equipment safety is combined with integrated management of intelligent operation and maintenance during the whole life cycle.

Meanwhile, in order to solve these new changes and new demands, a series of urgent scientific and technical problems have emerged, which will become scientific difficulties and technical bottlenecks in the development trend of smart ports, mainly including the following topics.

(1) Fusion and collaboration principle of material flow, energy flow, information flow of automatic equipment system.

(2) Intelligent planning and scheduling theory for the whole process (temporal and spatial) of autonomous handling systems.

(3) "Self-consistency" in the operation control of intelligent equipment.

(4) Intelligent design and realization technology of large port equipment.

(5) Design and realization technology of intelligent cooperative control system for handling facilities.

(6) Operation emulation, prediction analysis and realization technology based on big data.

The research and solution of these scientific and technological problems will effectively promote the construction and development of smart port. Meanwhile, the effective solution of these scientific and technological problems also depends on the application and support of high and new technologies.

11.3 Development Target of Smart Port

In the process of innovative application of these high and new technologies, intelligence of smart port infrastructure, data integration of logistics chain, intelligent operation and management, trade service facilitation of supply chain and ecological innovation and sharing have become the main development targets of smart ports in China.

1. Intellectualization of smart port infrastructure

Based on the cyber-physical system, interconnection among the environment, facilities, equipment and applications will be realized in the port area, the upgrade of cloud facilities will be accelerated, the construction of 5G and other network infrastructure will be strengthened, and advanced technologies such as the IOT will be applied to realize the comprehensive perception of facilities and equipment.

In terms of 5G network infrastructure, China will actively promote the construction of 5G network facilities, focusing on supporting the construction of 5G private networks in ports and logistics parks, and realize stable applications in large handling facilities, remote control of autonomous container trucks, and transmission of high-definition monitoring video, so that the intelligent container terminal can realize the application of 5G technology in full scenarios.

For the comprehensive perception of facilities and equipment, the industrial IOT and advanced sensor technology are applied to accelerate the digital upgrading of main facilities and equipment in the port, including promoting the tracking of core handling facilities and GPS vehicles, realizing the remote temperature monitoring system of reefer containers, realizing the remote supervision of dangerous goods containers (gas leaking, automatic spray), realizing real-time digital monitoring of energy consumption, etc.

The cloud service capacity will be strengthened, the operation, maintenance and management level of the cloud computing center of the smart port will be improved, and the software operation and maintenance service capacity of the cloud computing center will be further improved on the basis of stable hardware operation and maintenance. Relying on the digital middle-office, the intensive control advantages of the integrated cloud computing architecture of cloud computing will be fully explored. With the international first-class cloud computing service system and service standard as benchmark, the level of computing management, resource management, warning management, topology management, capacity management and configuration management will be comprehensively improved. Installation and maintenance services of all kinds of server operating systems, database systems, software tools and other services covering the whole process will be provided, so as to finally form the software service system of the port cloud data center.

2. Data integration of logistics chain

Data are taken as the core resource of the enterprise. The data integration of the port and shipping industry is divided into two levels, data resource acquisition and data integration application. The port and shipping data resource acquisition ability is deeply extended by relying on the construction of big data center, and the concept and technical framework of the digital middle-office are introduced to effectively promote the data integration and application of the logistics chain.

Relying on the construction of big data center, it will improve the big data service capability of the port and shipping industry, realize the exchange of data with the customs, maritime supervision and other government regulatory authorities, accumulate and build big data platform of the port and shipping, provide unified and open big data value-added services for logistics chain enterprises, and gradually form comprehensive big data service capability of port and shipping. The data acquisition of the whole process of port and shipping logistics will be strengthened, the online analysis and application based on the big data of operation will be realized to promote the evaluation and optimization of the intelligent operation of container terminals based on the big data analysis, and build a data-driven intelligent brain of ports. Data integration application of ship-shore logistics will be realized. By opening the data interaction channel between the pre-stowage center of shipping company/shipping agency and the TOS at the port, the real-time communication can be realized between the data of the pre-stowage center of the shipping company/agency and the data of container delivery and collection of the TOS of the terminal, so as to effectively improve the quality of ship stowage and improve the efficiency of handling operations of the terminal.

Based on cloud computing platform, the large middle-office system of port will be constructed, middle-office data services, middle-office operation services, middle-office authorized service ability of the digital port will be improved, integration service mode of modular system will be created, the construction of middle-office system of the GIS in the port, government regulation and other business will be promoted, the effective and efficient integration and utilization of port big data will be pushed, to construct the ecological digital system, and achieve a high degree of synergy in port operations and deep resource aggregation. Standardized data sharing services based on the middle-office platform architecture will be realized, Web API interface services of related data will be developed, data authorization and management system will be established, to provide big data services for logistics chain enterprises. The construction of the business middle-office that provides basic port geographic information services will be promoted, a unified high-precision geographic information system for smart ports will be established, to realize visualized interaction of electronic maps and 3D models, and implement applications and services in the fields of land, planning, facilities and equipment, etc.

3. Intelligent operation and management

To comply with the new development trend of port refinement, agility, flexibility and intellectualization, the deep integration of high and new technology with various fields in the port will be accelerated, the automation level of hardware equipment and the intellectualization level of software system of port operation will be comprehensively improved, to effectively enhance the intensive control ability, and achieve the port operation management of world-class level.

The China model of smart port and China plan of intelligent terminal have been put forward and implemented, adhering to the five concepts of "innovative, green, intelligent, secure and efficient", the remote control, remote monitoring and intelligent diagnosis system based on the new generation of communication network technology will be developed. Relying on modern information technologies such as AI, big data and cloud computing, the core systems with independent intellectual property rights such as intelligent scheduling, intelligent operation, intelligent traffic, intelligent gate, intelligent tally, intelligent energy, intelligent monitoring and intelligent security have been developed. Integrated innovation in key technologies for the construction and operation of new smart container terminals has been promoted, to build a container terminal with the highest degree of intelligence, the lowest construction cost and the best operation efficiency in the world.

The edge computing technology is applied to substantially increase the proportion of automated operation of hardware equipment. Automation technology for core handling facilities of container terminals is constantly developed, to improve the operating proportion of automatic equipment of container terminals. The intelligent upgrading of hardware equipment such as the automatic lead-sealing machine in the rear yard and customer self-service terminals is accelerated, so as to reduce personnel and improve efficiency. The scale application of iAGVs and autonomous container trucks will be gradually promoted, core systems oriented to scale application of autonomous container trucks, such as fleet management, vehicle scheduling, vehicle and road collaboration, intelligent transportation, will be developed, to break through the key techniques of the world leading level, such as the task plan management, precise positioning, real-time perception of road condition, automatic optimization of route, roadside collaborative management. The automation and intellectualization of dry bulk terminal equipment will be promoted, including the automation and intellectualization upgrading of core facilities such as ship loaders of dry bulk terminal, bucket ship unloaders, stacker-reclaimers. The automatic control technology of bucket ship unloaders and loaders in dry bulk terminal will be explored and studied, to realize the remote control of core facilities such as stacker-reclaimers, and maintain the world leading position in the automatic operation of dry bulk terminal.

The intellectualization level of software systems will be improved, and the research and localized development of port core software systems will be launched and accelerated. Aiming at the software system of container sector, the intelligent port TOS system and its intelligent control core module are developed with the goal of having the ability to replace import systems. Applying cutting-edge technologies and advanced architectures such as digital snapshot and digital twinning, the container terminal operation emulation system has been developed to realize the predictive deduction of container terminal handling operations and improve the pre-control level of operation logistics. The realization of lean management objectives such as operation plan evaluation, logistics bottleneck pre-inspection, operation time prediction, etc., will be promoted, to realize the digital holographic backtracking of operation processes, and initially acquire the ability to predict single ship operation. The intelligent upgrading of dry bulk terminal operation systems will be promoted. The subsystem of intelligent storage yard of dry bulk terminals will be studied and developed, to realize the intelligent storage yard plan, promote the research and development of intelligent modules such as berth scheduling and transportation capacity scheduling of dry bulk terminals, and maintain the advanced level of intelligent management of dry bulk terminals in China. The intelligent logistics integration system is upgraded. Aiming at the logistics sector, an integrated storage yard operation control system with unified, advanced structure and complete functions is established, and a professional intelligent control system for the fleet is built, so as to realize intelligent management modes such as intelligent vehicle monitoring, intelligent scheduling, intelligent settlement and intelligent service.

4. Trade service facilitation of supply chain

(1) The port business environment will be constantly improved. The paperless/electronic documents will be vigorously promoted, the online and offline collaborative service capacity will be improved, a developed hinterland transport network will be built, to improve hinterland logistics and trade service facilitation, then further promote the business coordination of the customs and the port, and provide customers with more valuable services of high-quality.

(2) Paperless documents in port will be comprehensively promoted. On the basis of the promotion and application of the existing container electronic documents system, a comprehensive application platform for electronic documents (including dry bulk electronic documents) will be further developed and established, so as to realize paperless business documents of the whole process between ports and shipping companies, shipping agencies, storage yards and terminals. The application of blockchain technology will be actively explored in the circulation process of container electronic delivery order to improve the compliance and security of electronic property certificates. With the whole process of electronic bill of lading based on blockchain technology as the breakthrough point, the paperless application of bill of lading in shipping will be accelerated by combining with large shipping companies, shipping agencies and other representative companies in the industry.

(3) Online and offline integrated services will be improved. The construction of a comprehensive online service platform for all kinds of port businesses will be comprehensively promoted, unifying data standards and service standards for all business sectors, and creating an integrated new port service mode that combines online services with offline services, to effectively improve customer service facilitation, and provide customers with 24-h online and offline services.

(4) An integrated control system for waterless ports will be established. Relying on the extensive layout of the waterless port, the developed whole-process supply chain network and hinterland transport network will be constructed, to improve the service facilitation for the logistics and trade in the waterless port area. The common needs of operation control, production management, logistics service and settlement system of the waterless port storage yard will be studied and refined, to design the intelligent management architecture of the waterless port based on the digital middle-office mode, and create a new,

cloud-based, standardized and highly configurable integrated management and control and comprehensive service system of the waterless port. Multimodal transport logistics system will be realized. Relying on the hinterland of the waterless port and the multimodal transport project, the research and development of intelligent multimodal transport unit and automatic handling and transshipment equipment will be promoted, research on remote monitoring, tracking and management system of container supply chain for the One Belt and One Road as well as international logistics and trade ecosystem for the One Belt and One Road will be implemented. A multimodal logistics comprehensive service platform will be established, to form the service capabilities of the professional logistics solutions, including online inquiry, online consultation, online transaction, online settlement and other service, as well as an "online plus offline" comprehensive logistics service network, and build a multimodal logistics comprehensive service platform within the scope of port.

(5) Business cooperation with government regulatory departments will be promoted. The business collaboration, data integration, process innovation, model innovation of the port logistics system with customs, maritime and other government regulatory departments will be actively promoted. Relying on the communication network architecture of the intelligent logistics system and its construction technology, a smart platform for collecting and delivering port goods will be constructed, to actively promote operation modes such as "direct delivery by ship", "direct loading after arrival at port", with reengineering of business processes, such as port handling, vehicle management and vehicle and cargo matching, as the core, an intelligent vehicle–road collaborative platform for port logistics visual collecting and delivering system will be established. By integrating social vehicle resources and optimizing vehicle scheduling, online booking, business matchmaking, online payment, route tracking and other functions of container collecting and delivering business will be realized, to create an innovative new mode of business cooperation between customs and port. In order to effectively improve the business collaboration between ports and maritime departments, an innovative and intelligent maritime service platform will be built by sorting out the collaborative needs of maritime business and port business, formulating the standards for data fusion and business collaboration of maritime ports, and thus effectively improving the level of business collaboration between ports and maritime departments.

5. Ecological innovation sharing

A new concept of open, sharing and win–win cooperation should be fostered, give full play to the leading role of Shanghai, Tianjin, Shenzhen and other hub ports in the port industry in China, and build a smart port ecological system featuring port data and information hubs, innovative service platforms and knowledge sharing, so as to effectively boost the construction of smart ports.

The innovative application of new technologies should be strengthened. 5G, IPv6, Beidou navigation and other cutting-edge information technologies should be closely tracked, exchanges and cooperation with domestic and foreign professional institutions, scientific research institutes and industry-leading enterprises should be strengthened, the latest application trends should be actively grasped, and the comprehensive application of emerging technologies in the port should be accelerated. Key projects, demonstration projects and positioning technology actively adopting the sub-meter Beidou of high-precision should be encouraged, and prominent representative problems in the industry, such as signal shielding under large equipment should be broken through with practical application.

We will lead industry standard research and development. We will formulate or participate in making a series of organization standards, industry standards, national standards and international standards related to smart ports in automated terminals, new technology and technique requirements, data and interface specification, to cultivate professional talents in the field of smart port standards, actively grasp leading right for standards formulation, and fully play the leading role in the smart port industry.

The evaluation system of smart ports should be explored. Relying on the existing achievements of smart port construction and major projects that are representative of the industry, the core quantitative indicators of next-generation smart container terminals and smart dry bulk cargo terminals will be studied, and the evaluation system of smart port construction and development will be established, laying a foundation for the research and release of smart port index.

In the background of the new round of information technology revolution, the "unmanned" and "intelligent" operation and service of the port will become the main form of presentation. In general, by 2025, all the major hub ports in China will be built into smart container ports, and by 2030, all the major hub ports in China will be built into world-class smart ports, and an intelligent collection, delivery and transportation logistics chain system with smart ports as the core will be formed, leading the new development direction of the world's smart ports.

Printed in the United States
by Baker & Taylor Publisher Services